DISTILLATION CONTROL

DISTILLATION CONTROL

An Engineering Perspective

CECIL L. SMITH

WILEY

A JOHN WILEY & SONS, INC., PUBLICATION

Published by John Wiley & Sons, Inc., Hoboken, New Jersey
Published simultaneously in Canada

For general information on our other products and services or for technical support, please contact our Customer Care Department within the United States at (800) 762-2974, outside the United States at (317) 572-3993 or fax (317) 572-4002.

Wiley also publishes its books in a variety of electronic formats. Some content that appears in print may not be available in electronic formats. For more information about Wiley products, visit our web site at www.wiley.com.

Library of Congress Cataloging-in-Publication Data:

Smith, Cecil L.
    Distillation control : an engineering perspective / Cecil L. Smith.
        p. cm.
    Includes bibliographical references and index.
    ISBN 978-0-470-38194-6
    1. Distillation.   I. Title.
    TP156.D5D586 2012
    660'.28425–dc23

                                                            2011041437

ISBN: 9780470381946

CONTENTS

CONTENTS

PREFACE

Two observations constitute the basis for this book:

1. Despite its thirst for energy, distillation continues to be widely used for separations. Efficiently operating these columns requires a high degree of automatic control.
2. Virtually all column designs are based on a steady-state separation model. Especially for columns separating nonideal materials, there is no alternative.

The perspective of this book is that the steady-state separation model should also be the basis for developing the control configuration for the column. Yes, a steady-state model! Although the technology to do so is widely available, extending to a dynamic model is not necessary for developing the column control configuration.

The most crucial component of every process control application is developing the piping and instrumentation (P&I) diagram that defines the control configuration for the process and for each unit operation, such as distillation, within that process. If the P&I diagram is correct, the loops can be successfully commissioned and tuned to deliver the required performance. But where the configuration is deficient, the usual consequence is tuning difficulties. Until the deficiencies in the P&I diagram are corrected, neither automatic tuning, tuning techniques, nor experienced tuning professionals can succeed.

For something so crucial to success in process control, one would think rigorous procedures would be available to derive the P&I diagram from the process characteristics, operating objectives, and so on. Instead, the usual practice is basically copying—the control configuration from a sister plant with the same or similar process is used as the starting point for the P&I diagram. This works reasonably well in power generation, pulp and paper, oil refining, and other industries where the same basic process technology is being replicated, but with different production rates, different feedstocks, and so forth. How many outright mistakes have been copied? How many times has a poorly performing configuration been copied when a better performing configuration could be implemented? Despite an occasional "war story," the answers to such questions are largely opinions.

One should expect better, specifically, a rigorous procedure for translating the characteristics of the process (as expressed by models) and the operating objectives into a P&I diagram. This would also be useful when choosing between design alternatives, thus promoting the integration of process design and process control. Steady-state models are now available for all unit operations, and such models are the basis for most modern plant designs. Especially for continuous processes, the process flow sheet is developed using these models. Such models should also provide the basis for developing the P&I diagram.

For too long, the primary focus of process control has been the linear systems theory. Rarely is such technology useful in developing a P&I diagram. This perspective is the basis of another misconception, specifically, that the dynamic behavior of the process dictates the appropriate control configuration. This seems to translate to "control every variable with the nearest valve" as the guiding principle for developing a P&I diagram. Is this done consciously? Not usually, but if you examine enough P&I diagrams, it seems to turn out that way. However, if process dynamics receive the primary consideration in developing the control configuration, this would often translate to "control every variable with the nearest valve."

The steady-state characteristics of the process largely determine the appropriate control configuration. What is the direct and long-term influence of a final control element on one or more controlled variables? When developing a P&I diagram, the customary practice is to rely on a qualitative assessment. While this is often sufficient, processes can be subtle and occasionally behave very differently from what is expected. When this occurs, the resulting P&I diagram is deficient. This prospect increases with the complexity of the process, with the haste with which the P&I diagram must be developed, and with the inexperience of the developer of the P&I diagram.

Process characteristics are best expressed in the form of a model for the process. Given the current availability of such models, it is time to begin relying on a quantitative assessment of process characteristics. This is short of the ultimate goal, namely to derive the P&I diagram from such models. However, this is a step in the right direction, and distillation is a good unit operation to use as the starting point. Operating variables such as product flows, reflux, and boilup affect the composition of all product streams, but not to the same degree. The selection of the control configuration is preferably based on a quantitative assessment of their effect. For this, the steady-state separation model suffices.

Single-end composition control is rather forgiving. Double-end composition control is not. The same can be said for sidestream towers for which two product compositions must be controlled. For columns separating well-behaved materials, statements can be developed to guide the choice of the control configuration. However, these statements must be used cautiously for columns separating nonideal materials. In either case, the preferable approach is to base the choice of the control configuration on a quantitative assessment of column behavior computed from the steady-state separation model used for column design.

Houston, Texas   CECIL L. SMITH
November 28, 2011

1

PRINCIPLES

A distillation column obtains separation through energy. Consequently, it seems intuitive that a product composition must be controlled by manipulating a term relating to energy. When the composition of both product streams from a two-product tower must be controlled, this suggests the following approach:

- Control the distillate composition by adjusting the reflux.
- Control the bottoms composition by adjusting the boilup.

For most columns, this control configuration exhibits a substantial degree of interaction, which translates to operational problems in the field.

An alternate approach is as follows:

- Control the composition of one of the products (distillate or bottoms) by adjusting an energy term (reflux or boilup).
- Control the composition of the other product by adjusting the respective product draw.

For most applications, the degree of interaction is much lower.

With this approach, one of the compositions is being controlled by directly adjusting a term in the column material balance. Consequently, this presentation begins with various material balances (entire tower, condenser only,

---

Distillation Control: An Engineering Perspective, First Edition. Cecil L. Smith.
© 2012 John Wiley & Sons, Inc. Published 2012 by John Wiley & Sons, Inc.

reboiler only). The discussion proceeds to component material balances for binary distillation, followed by an examination of the relationship between energy and separation. The primary objective is to provide insight into the nature of distillation and make the case that controlling one of the product compositions by adjusting a product draw is not only possible but is likely to be the appropriate approach for most towers.

This chapter reviews the general principles of distillation that are relevant to process control, including

- material balances, energy, and separation;
- composition control, through either energy terms or product flows;
- the stage-by-stage separation models for multicomponent distillation and their utility in control analyses;
- tray towers and packed towers;
- column dynamics.

1.1. SEPARATION PROCESSES

A simple separation process splits a feed stream into two product steams. In a pure separation process, no molecules are created, rearranged, or destroyed. That is, every molecule in the feed stream appears unchanged in one of the product streams.

Examples of industrial separation processes include the following:

- adsorbers
- centrifuges
- crystallizers
- cyclones
- decanters
- distillation columns
- dryers
- evaporators
- filters
- mist extractors

Every separation process relies on some principle to separate the molecules. Some separate by phases—a filter separates solids from liquids, a mist extractor separates liquids from gases, a decanter separates two immiscible liquids. Some separate by forcing a phase change—a dryer vaporizes a component such as water, leaving the nonvolatile solids behind. Distillation separates components based on their differences in volatility.

Separation processes, and distillation in particular, can become quite complex. Multiple feeds are possible. Multiple product streams are very common in distillation applications. Considerations such as energy conservation often add complexity to improve overall energy efficiency. Even reactive distillation systems are now occasionally incorporated into plant designs.

1.1.1. Binary Distillation

A binary separation process is one for which the feed contains only two components. Most presentations begin with such processes, as they are the simplest cases. Binary separations are occasionally encountered in practice, but most industrial columns are multicomponent.

A binary distillation example commonly used in textbooks is a column whose feed is a mixture of benzene and toluene. At atmospheric pressure, benzene boils at 80.1°C; toluene boils at 110.8°C. Consequently, benzene is more volatile than toluene. If a mixture of benzene and toluene is heated to its bubble point, the benzene vaporizes preferentially to the toluene. If the mixture is 50% benzene and 50% toluene, the vapor will contain more than 50% benzene and less than 50% toluene.

In distillation, the terms "light" and "heavy" are used to distinguish the components. But as used in distillation, these terms do not reflect weight, density, and so on. The light component is the more volatile; the heavy component is the less volatile. This notation is also reflected in the subscripts that designate the components:

x_L = mole fraction of light component in a liquid stream or phase;
x_H = mole fraction of heavy component in a liquid stream or phase;
y_L = mole fraction of light component in a vapor stream or phase;
y_H = mole fraction of heavy component in a vapor stream or phase.

1.1.2. Stages

A stage provides an arrangement where a vapor phase is in equilibrium with a liquid phase. The more volatile components concentrate in the vapor phase. The less volatile components concentrate in the liquid phase. The relationship between the vapor composition and the liquid composition is governed by the vapor–liquid equilibrium relationships for the various components.

A flash drum is a separation process that consists of a single stage. The feed is a superheated liquid that partially vaporizes (or flashes) within the flash drum. The two phases are separated to provide a vapor stream and a liquid stream. These are assumed to be in equilibrium as per the vapor–liquid equilibrium relationships.

Such single-stage separations are only viable when a crude separation is required between materials of significant difference in volatility. In distillation

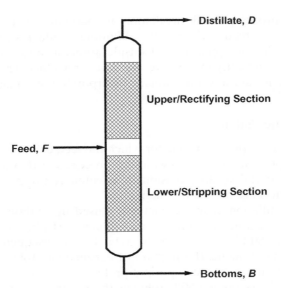

Figure 1.1. Distillation column.

columns, a separation section provides a sequence of stages whereby liquid flowing down the section is successively contacted with the vapor flowing up the section. One approach is to use trays to provide the vapor–liquid contact, with each tray ideally providing one stage (actual trays are not quite that good). The alternate approach is to use packing to provide the vapor–liquid contact. The selection of trays versus packing is a design issue with surprisingly little impact on the column controls.

As illustrated in Figure 1.1, a two-product tower contains two separation sections, one (the upper or rectifying section) between the feed and the distillate, and the other (the lower or stripping section) between the feed and the bottoms. The number of stages required in each section is determined by the design of the column. The controls have no way to influence the number of stages in each section.

Designs are usually based on "ideal stages," where the vapor and liquid on the stage are in equilibrium. Actual stages rarely achieve this. A parameter known as the stage efficiency quantifies the departure of a stage from ideality. This parameter is used to adjust the actual number of stages installed in a column.

1.1.3. Engineering Units

For operator displays, reports, and other indications in production operations, the engineering units are typically as follows:

Flows. Either mass flow (kg/h, lb/h, etc.) or volumetric flow (L/h, gal/h, etc.).

Compositions. Either weight percent (wt%) or volume percent (vol%) for liquids; usually vol% (= mol%) for gases and vapors.

However, vapor–liquid equilibrium relationships are fundamentally based on molar quantities. Consequently, the equations used for the design, analysis, and so on, of distillation columns are normally developed in molar units:

Flows. Molar flow (mol/h, mol/min, etc.).

Compositions. Mole fractions.

Herein molar units will generally be used for both flows and compositions.

1.1.4. Feed and Product Streams

Figure 1.1 illustrates a two-product distillation column with a single feed stream. The designation of the streams is usually as follows:

Feed. The flow rate of this stream will be designated by F, in mol/h.

Distillate. The flow rate of this stream will be designated by D, in mol/h. This stream is sometimes referred to as the overheads.

Bottoms. The flow rate of this stream will be designated by B, in mol/h.

Feed composition. The possibilities for the feed stream F are as follows:
- entirely liquid,
- entirely vapor,
- vapor–liquid mixture.

The mole fraction of such streams is normally designated by z. The composition of the light component is z_L; the composition of the heavy component is z_H.

1.1.5. Distillate Composition

The possibilities for the distillate stream are as follows:

Entirely liquid. The condenser must be a total condenser as illustrated in Figure 1.2a. The overhead vapor V_C that flows into the condenser is totally condensed to provide liquid for the distillate stream and the reflux stream. The composition of the distillate is the same as the composition of the overhead vapor.

Entirely vapor. The condenser must be a partial condenser as illustrated in Figure 1.2b. Only part of the overhead vapor V_C flowing into the condenser is condensed. The resulting liquid is the reflux stream. The

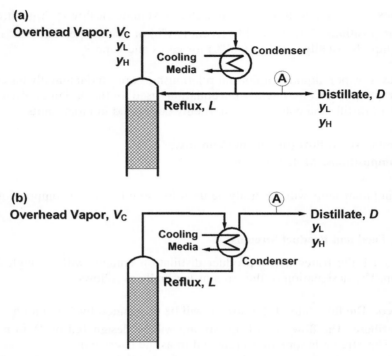

Figure 1.2. Overhead composition. (a) Total condenser. (b) Partial condenser.

distillate stream is the vapor that is not condensed. A partial condenser provides separation and is ideally one stage. The composition of the distillate is not the same as the composition of the overhead vapor.

The distillate composition is either the composition of a vapor stream (partial condenser) or the composition of a vapor stream that is condensed (total condenser) to provide the liquid overhead product. Vapor compositions are normally designated by y, giving the following notation for the distillate composition:

y_L = mole fraction of the light component;

y_H = mole fraction of the heavy component.

1.1.6. Bottoms Composition

As illustrated in Figure 1.3, the bottoms stream is always a liquid stream. Only part of the liquid flowing into the reboiler is vaporized, making the reboiler the counterpart of the partial condenser. The vapor stream becomes the boilup to the column; the liquid stream is the bottoms product.

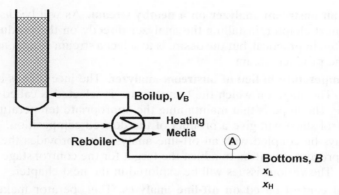

Figure 1.3. Bottoms composition.

Liquid compositions are normally designated by x, giving the following notation for the bottoms composition:

x_L = mole fraction of the light component;
x_H = mole fraction of the heavy component.

1.1.7. Composition Measurement

The performance of a column ultimately depends on the composition of the product streams. There are two possibilities:

Single-end composition control. The composition of one of the product streams is controlled, and the other is allowed to "float."

Double-end composition control. The composition of both product streams is controlled. This is far more challenging.

The specification for the composition of a product stream can be in many forms, some of which will be examined in the next chapter. Throughout this book, the composition of a product stream will be stated in terms of one or more impurities. For a binary separation, the only impurity in the distillate composition is y_H; the only impurity in the bottoms is x_L. The smaller the value of y_H, the higher the purity of the distillate product. The smaller the value of x_L, the higher the purity of the bottoms product.

Ideally, a product composition would be sensed by an onstream analyzer installed on the product stream, as is illustrated in Figures 1.2a,b and 1.3. This will be the general practice in the piping and instrumentation (P&I) diagrams presented in this book. But unfortunately, practical considerations often dictate otherwise, the options generally being the following:

Install an onstream analyzer on a nearby stream. As will be discussed in the next chapter, installing the analyzer directly on the product stream is often impractical, but the desire is to select a stream as near as possible to the product stream.

Use temperature in lieu of onstream analyzer. The incentive is obvious—cost. The stage on which the temperature is selected is called a *control stage*. The hope is that maintaining the appropriate temperature on the control stage will give a product of the desired composition. This must always be coupled with an off-line analysis that provides the basis for the process operators to adjust the target for the control stage temperature. The various issues will be explored in the next chapter.

Manual control based on off-line analyses. The operator makes adjustments based on the results of the off-line analyses. The downside of this approach is that the product compositions are conservatively maintained within specification, which results in reduced throughput, lower yields (loss of valuable product through a product stream), increased energy costs, and so on.

The P&I diagrams in this book will generally illustrate composition control based on a composition analyzer installed directly on a product stream. This is the ideal, and the closer it can be achieved in practice, the better.

1.1.8. Manipulated Variables

In distillation applications, the most common final control elements are control valves, although pumps with variable speed drives are certainly viable alternatives. Consequently, the output of most controllers will be a control valve opening. This valve opening in turn determines the flow through the control valve.

Technically, the manipulated variable would be the control valve opening. However, the various relationships (material balances, energy balances, etc.) that will be written for a column invariably involve flows, not valve openings. The variables in distillation simulation programs are always flows, never valve openings. Consequently, in this book, the flow through the control valve will be routinely referred to as the manipulated variable.

In older towers, flow measurements were rather sparingly installed. But in newer towers, flow measurements are more widely applied, and in some, a flow measurement is installed on every stream where metering is possible. The availability of a flow measurement permits a flow controller to be configured in the controls, and cascade control configured for loops such as composition and level. In cascade control, the output of the outer loop (composition, level, etc.) is the set point of the inner loop (flow). Technically, the manipulated variable for the outer loop is a flow set point, but as flow controllers are far faster than composition, level, and so on, the actual flow is essentially equal to its set point, at least from the perspective of the slower loop. In the cascade configurations, the manipulated variable for the outer loop is essentially a flow.

As composition loops are very slow, providing a flow controller as an inner loop is generally recommended. In this book, cascade will be indicated for composition loops and for temperature loops for the upper and lower control stages. For level loops, providing a flow controller for the inner loop is not essential, especially when close control of level is not required. Within this book, cascade control will not generally be configured for level loops. However, if a flow measurement is available for other reasons, cascade control should be configured in practice.

1.2. TOTAL MATERIAL BALANCE

Material balances are the most fundamental equations that can be written for any process. For the two-product distillation column illustrated in Figure 1.4, the steady-state total material balance is written as follows:

$$F = D + B.$$

On a long-term basis, this equation must close. If the feed flow is constant, then

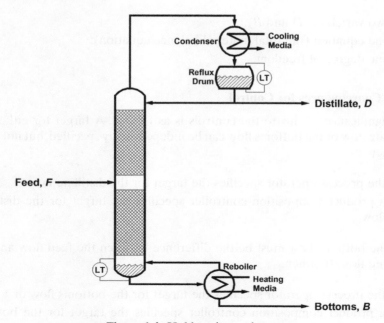

Figure 1.4. Holdups in a column.

1. any long-term change in the distillate flow must be offset by an equal and opposite change in the bottoms flow;
2. any long-term change in the bottoms flow must be offset by an equal and opposite change in the distillate flow.

1.2.1. Degrees of Freedom

The control configuration must be consistent with the degrees of freedom for the process. The equation for the degrees of freedom is as follows:

Degrees of freedom = number of variables – number of equations.

Most distillation columns are said to operate in a "fixed service," which means that

1. the feed flow F is explicitly specified or is determined by upstream unit operations;
2. the feed composition is determined by upstream unit operations.

In such columns, the feed flow F is considered to be a known quantity in the material balance equation. This leaves two variables in the material balance equation, specifically, the distillate flow D and the bottoms flow B. Therefore, there are

- two variables (D and B);
- one equation (the total material balance equation);
- one degree of freedom.

1.2.2. Consequences for Control

The significance of this to the controls is as follows. A target for either the distillate flow or the bottoms flow can be independently specified, but not both. If either

1. the process operator specifies the target for the distillate flow or
2. a product composition controller specifies the target for the distillate flow,

then the bottoms flow must be the difference between the feed flow and the distillate flow. If either

1. the process operator specifies the target for the bottoms flow or
2. a product composition controller specifies the target for the bottoms flow,

then the distillate flow must be the difference between the feed flow and the bottoms flow.

1.2.3. Unsteady-State Behavior

At unsteady state, the possibilities are as follows:

1. Feed rate exceeds the sum of the product rates. Material accumulates somewhere within the tower.
2. Feed rate is less than the sum of the product rates. Material depletes somewhere within the tower.

Material accumulates or depletes primarily either in the reflux drum, in the bottom of the column, or both.

The amount of material (holdup) on the tower internals (trays or packing) is not constant. However, this holdup is largely determined by the design of the internals. The internal flows (reflux and boilup) have some influence on this holdup. However, the product flows (distillate and bottoms) have no direct influence on this holdup. Any long-term imbalance in the steady-state material balance will affect the holdup in the reflux drum and/or in the bottoms of the tower.

1.2.4. Level Measurement

As illustrated in Figure 1.4, level measurements are normally provided on both holdups. The capacity of these holdups is limited by the size of the equipment, so high and low level switches are usually installed in the reflux drum and in the bottoms. So that these switches are not actuated, one responsibility of the control configuration is to force the closure of the overall material balance by maintaining the levels within a "reasonable proximity" of their targets.

A level measurement for the bottoms holdup is essentially universal, but for the condenser, there are exceptions:

Flooded condenser. The condenser is partially filled with liquid, which reduces the effective area for condensing the overhead vapor. The level within the condenser is allowed to seek its own equilibrium, which means that sufficient heat transfer area is exposed to condense the overhead vapor. The level is never controlled and usually not measured.

No reflux drum. In small-diameter towers that require an external structure for support, the condenser is often physically mounted on the top of the tower. The reflux is returned directly to the tower, so no reflux drum is required.

These will be discussed in more detail in the subsequent chapter devoted to condenser arrangements.

1.2.5. Integrating Process

Consider the behavior of the process under the following conditions:

1. Process is within its design limits (no vessel capacities exceeded; no vessel empty).
2. No controls are on automatic.

Let H be the total holdup of material within the column. Changes in holdup affect the head for fluid flow. This is significant only for gravity flow applications, which are rare in distillation. Otherwise, changes in the holdup H have no direct effect on either the feed flow F, the distillate flow D, or the bottoms flow B.

The unsteady-state material balance can be written in either its differential or its integrated form:

$$\text{Differential: } \frac{dH(t)}{dt} = F(t) - D(t) - B(t)$$

$$\text{Integrated: } H(t) = \int [F(t) - D(t) - B(t)] dt$$

When H has no effect on F, D, or B, a process described by such equations is referred to as an *integrating process*. An alternate term is *ramp process* (the response to any upset is a ramp in the holdup or level) or *non-self-regulated process* (the process will not seek an equilibrium unless control actions are taken).

1.2.6. Level Control

An integrating process does not seek its own equilibrium. If there is an imbalance in the total material balance, the result is one of the following:

$F > B + D$. The holdup increases until some limiting condition is attained, the limiting condition being either
    1. the level in the reflux drum actuates the high level switch or
    2. the level in the bottoms actuates the high level switch.
$F < B + D$. The holdup decreases until some limiting condition is attained, the limiting condition being either
    1. the level in the reflux drum actuates the low level switch or
    2. the level in the bottoms actuates the low level switch.

The responsibility of every level controller is to close some material balance. To assure that the column material balance closes, every column control configuration must contain one of the following:

1. The reflux drum level is controlled by manipulating the distillate flow.
2. The bottoms level is controlled by manipulating the bottoms flow.

Providing both is also an option.

1.3. REFLUX AND BOILUP RATIOS

The reflux L and boilup V are associated with energy. The heat supplied to the reboiler generates the boilup V. In a partial condenser (distillate product is a vapor stream), the heat removed by the condenser generates the reflux L. In this context, several ratios arise, most of which involve the ratio of a liquid flow and a vapor flow.

1.3.1. External Reflux Ratio

The external reflux ratio is the ratio of the reflux flow L to the distillate flow D:

$$\text{External reflux ratio} = \frac{L}{D}.$$

In many towers, flow measurements can be installed for these two flows, and if so, the external reflux ratio can be computed.

However, there are tower designs where measurement of the reflux flow is not possible. To minimize pressure drops in vacuum towers, the condenser is often physically mounted on the top of the column. For a partial condenser, all of the condensate is returned directly to the column to provide the reflux. For a total condenser, part of the condensate is withdrawn with the remainder returned directly to the column to provide the reflux. In neither arrangement is it possible to measure the reflux flow.

1.3.2. Boilup Ratio

The counterpart to the external reflux ratio (that pertains to the top of the tower) is the boilup ratio (which pertains to the bottom of the tower). The boilup ratio is the ratio of the boilup V to the bottoms flow B:

$$\text{Boilup ratio} = \frac{V}{B}.$$

Direct measurement of the boilup flow V is never possible. Therefore, the boilup ratio cannot be computed from direct flow measurements.

When sufficient measurements are available to compute the energy transferred from the heating media to the reboiler, the boilup can be estimated by dividing this heat transfer rate by the latent heat of vaporization of the material in the reboiler. The simplest case is a steam-heated reboiler with a measurement for the steam flow S. The boilup V can be computed as follows:

$$V \cong \frac{S \cdot \lambda_S}{\lambda_B},$$

where

    λ_B = latent heat of vaporization of liquid in the reboiler;
    λ_S = latent heat of vaporization of the steam.

Unfortunately, there is always some error in the resulting value.

If the objective is to maintain a constant boilup flow, one possibility is to measure the pressure drop across a few of the lower stages and adjust the heat to the reboiler to maintain a constant pressure drop. One must use enough stages so that the pressure drop being sensed is above the noise invariably associated with such measurements. Furthermore, the pressure drop is related to the square of the vapor flow, so this approach works better at high vapor flows than at low vapor flows.

1.3.3. Internal Reflux Ratio

The internal reflux ratio R_I is the ratio of the reflux flow L to the vapor flow V at a point within the tower:

$$R_{I,k} = \frac{L_k}{V_k},$$

where

    L_k = reflux flow at location k within the tower;
    V_k = vapor flow at location k within the tower;
    $R_{I,k}$ = internal reflux ratio at location k within the tower.

The vapor and liquid flows within most columns vary from stage to stage, so the internal reflux ratio is not constant. Furthermore, the internal reflux ratio above the feed stage will be different from the internal reflux ratio below the feed stage.

1.3.4. Above Feed Stage

For a location above the feed stage, Figure 1.5 presents the streams for a total material balance from that location through the top of the column. The total material balance is as follows:

$$V_k - L_k = D.$$

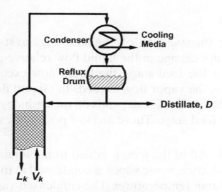

Figure 1.5. Internal reflux ratio above the feed stage.

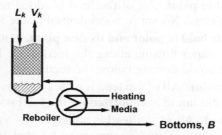

Figure 1.6. Internal reflux ratio below the feed stage.

Since the distillate flow D cannot be negative, the following conclusions can be made for the flows above the feed stage:

$$V_k \geq L_k,$$
$$R_{I,k} \leq 1.$$

1.3.5. Below Feed Stage

For a location below the feed stage, Figure 1.6 presents the streams for a total material balance from that location through the bottom of the column. The total material balance is as follows:

$$L_k - V_k = B.$$

Since the bottoms flow B cannot be negative, the following conclusions can be made for the flows below the feed stage:

$$L_k \geq V_k,$$
$$R_{I,k} \geq 1.$$

1.3.6. At Feed Stage

If one proceeds from the stages below the feed stage to stages above the feed state, there is an abrupt change in the liquid flow relative to the vapor flow at the feed stage. Below the feed stage, the liquid flow exceeds the vapor flow. Above the feed stage, the vapor flow exceeds the liquid flow.

What happens at the feed stage depends on the enthalpy of the feed relative to conditions on the feed stage. There are five possibilities:

Feed is subcooled. All of the feed is added to the liquid flowing below the feed stage. In addition, some vapor is condensed at the feed stage to heat the feed to column temperatures. The condensed vapor is added to the liquid flowing below the feed stage, but is removed from the vapor flowing above the feed stage.

Feed is at its bubble point. All of the feed is added to the liquid flowing below the feed stage. No vapor is condensed at the feed stage.

Feed is between its bubble point and its dew point. Some feed flashes and is added to the vapor flowing above the feed stage. The remaining feed is added to the liquid flowing below the feed stage.

Feed is at its dew point. All of the feed is added to the vapor flowing above the feed stage. No liquid is vaporized on the feed stage.

Feed is superheated. All of the feed is added to the vapor flowing above the feed stage. Some liquid is vaporized to cool the feed to column temperatures. The vaporized liquid is added to the vapor flowing above the feed stage, but is removed from the liquid flowing below the feed stage.

Most process designs avoid highly subcooled feeds and highly superheated vapors.

1.3.7. Total Reflux

Most towers can be operated with the feed shut off and both product draws shut off. Sometimes this is during startup; sometimes this is during a temporary interruption in production operations.

If no distillate product is being withdrawn, all of the overhead vapor is condensed and returned to the column as reflux. The external reflux ratio is infinite, but the internal reflux ratio above the feed stage is exactly 1.0.

If no bottoms product is being withdrawn, all of the bottoms liquid is vaporized and returned to the column as boilup. The boilup ratio is infinite, but the internal reflux ratio below the feed stage is exactly 1.0.

At least theoretically, columns can operate indefinitely at total reflux. But in practice, total reflux is a temporary situation, although temporary could be hours or perhaps days. Energy is being consumed, but no product is being made—not a good mode of operation with regards to the profit and loss statement. Production personnel must weigh the costs of continuing operation at total reflux versus the cost of shutting the tower down and restarting it.

1.3.8. Equimolal Overflow

On every stage within a separation section, some vapor is condensed and some liquid is vaporized. Equimolal overflow means that for each mole of vapor that is condensed, exactly one mole of liquid is vaporized. This is definitely not assured. Separations involving light hydrocarbons (ethane, propane, etc.) deviate less than separations involving more complex components.

When equimolal overflow is assumed, the liquid and vapor flows within a separation section do not change from stage to stage. The liquid flow on all stages within the upper separation section is the reflux L. The vapor flow on all stages within the lower separation section is the boilup V.

At the feed stage, there will be a change in the liquid and/or vapor flows. One way to characterize the enthalpy of the feed is by its quality q, which is the fraction of the feed that vaporizes at the feed stage. The value of q for various types of feed is as follows:

$q < 0$. Subcooled feed; some vapor is condensed at the feed stage to heat the feed to column temperatures.

$q = 0$. Liquid feed at its bubble point; none of the feed is vaporized.

$0 < q < 1$. Partially vaporized feed.

$q = 1$. Vapor feed at its dew point; none of the feed is condensed.

$q > 1$. Feed is a superheated vapor; some liquid is vaporized at the feed stage to cool the feed to column temperatures.

When equimolal overflow is assumed, the liquid flow L_B in the lower separation section is computed as follows:

$$L_B = L + (1-q)\,F.$$

The vapor flow throughout the upper separation section is the same as the overhead vapor flow V_C into the condenser and is computed as follows:

$$V_C = V + q\,F.$$

The assumption of equimolal overflow permits the liquid and vapor flows throughout the column to be easily computed. However, the results are approximate. For some separations, the liquid and vapor flows within a separation section change by a factor of 2 or more.

1.4. TOTAL MATERIAL BALANCE AROUND CONDENSER

A subsequent chapter is devoted to the wide variety of possible condenser configurations. A mechanism to influence the heat removed in the condenser is required, but the exact nature of this mechanism has no effect on the discussion that follows. The illustrations will only show a generic "cooling media" for a total condenser, but the discussion herein also applies to a partial condenser.

For small-diameter towers that require a structure for support, the condenser and reflux drum are usually physically located at the top of the column. But for a tower whose diameter is large enough that a structure is not required for support, cost issues favor the following configuration:

- The overhead vapor line extends to grade level.
- The condenser and reflux drum are physically at grade level.
- A reflux pump is required to return the reflux to the top stage.

No control issues are associated with any of this, so this detail will not be included in any of the illustrations in this book.

1.4.1. Condenser Material Balance

In the context of the material balance, the term "condenser" also includes the reflux drum, if one is present. The material balance contains a term for each of the three streams illustrated in Figure 1.7:

Distillate D (an output term). This is one of the product streams from the column. The controls influence the distillate flow via a control valve on the distillate stream.

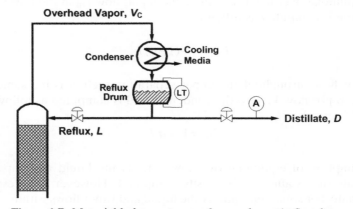

Figure 1.7. Material balance streams for condenser/reflux drum.

Reflux L (an output term). Part of the overhead vapor must be returned to the column as a liquid stream known as reflux. In most columns, the controls influence the reflux flow via a control valve on the reflux stream.

Overhead vapor V_C (an input term). This is determined by the heat removed in the condenser, which for most total condensers is adjusted by the tower pressure controller to maintain constant tower pressure. The material balance controls at the top of the column have no way to influence the overhead vapor flow.

The unsteady-state material balance around the condenser is written as follows:

$$V_C - (D + L) = \frac{dH_C}{dt},$$

where H_C is the reflux drum holdup (mole).

1.4.2. Control Configurations

The two manipulated variables, the distillate flow D and the reflux flow L, associated with the condenser are used to control the following two variables:

Distillate composition. When the distillate product is a salable product, good distillate composition control is crucial.

Reflux drum level. Rarely does the drum level affect any term in the profit-and-loss statement.

In selecting the control configuration, controlling the distillate composition must take priority, as reflected in the following approach:

1. Determine if the distillate composition is to be controlled by manipulating the reflux flow L or by manipulating the distillate flow D. This takes precedence over the usual preference to control level by manipulating the larger of the two flows (D or L).
2. Control reflux drum level with the other flow. However, level cannot be controlled by manipulating a very small flow. If $L/D \ll 1$, drum level cannot be controlled by manipulating L. If $L/D \gg 1$, drum level cannot be controlled by manipulating D.

Figure 1.8 presents the two possible control configurations, which are designated *direct material balance control* and *indirect material balance control*. The distillate flow D appears explicitly in the total material balance for the column:

$$F = D + B.$$

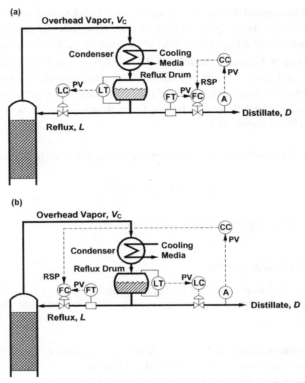

Figure 1.8. Control configurations for distillate composition. (a) Direct material balance control. (b) Indirect material balance control.

TABLE 1.1. Control Configurations for Distillate Composition

	Direct Material Balance Control	Indirect Material Balance Control
Control configuration	Figure 1.8a	Figure 1.8b
Manipulated variable for composition	Distillate D	Reflux L
Manipulated variable for drum level	Reflux L	Distillate D
Solution of condenser material balance	$L = V_C - D$	$D = V_C - L$
Preferred for level control if	$L > D$	$D > L$
Impractical if	$L/D \ll 1$	$L/D \gg 1$

The terms direct material balance control and indirect material balance control pertain to how the value of the distillate flow is determined. Table 1.1 summarizes the attributes of the two configurations.

The configuration in Figure 1.8a is the direct material balance control configuration. Values for D and L are determined as follows:

D—specified by the distillate composition controller;

L—determined by the level controller to satisfy the steady-state material balance for the condenser:

$$L = V_C - D.$$

The manipulated variable D for the composition controller appears explicitly in the column material balance.

The configuration in Figure 1.8b is the indirect material balance control configuration. Values for D and L are determined as follows:

L—specified by the distillate composition controller;

D—determined by the level controller to satisfy the steady-state material balance for the condenser:

$$D = V_C - L.$$

The manipulated variable L for the composition controller does not appear explicitly in the column material balance. Instead, the composition controller specifies L, from which the level controller determines the value of D.

1.5. TOTAL MATERIAL BALANCE AROUND REBOILER

A subsequent chapter is devoted to the wide variety of possible arrangements for reboilers at the bottom of the column. A mechanism to influence the heat added in the reboiler is required, but the exact nature of this mechanism has no effect on the discussion that follows. The illustrations will be for a steam-heated reboiler with a control valve and possibly a flow controller on the steam supply.

1.5.1. Reboiler Material Balance

In the context of the material balance, the term "reboiler" also includes the bottoms holdup. In Figure 1.9, the holdup for bottoms liquid is within the tower itself, but for kettle reboilers, this is within the reboiler. The material balance contains a term for each of the three streams illustrated in Figure 1.9:

Bottoms B (an output term). This is one of the product streams from the column. The controls influence the bottoms flow via a control valve on the bottoms stream.

Boilup V (an output term). Part of the liquid leaving the lower separation section of the column must be returned to the column as a vapor stream

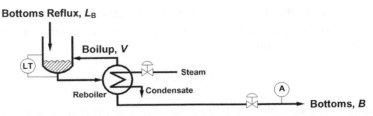

Figure 1.9. Material balance streams for reboiler.

known as boilup. Installing a control valve (or any other final control element) on the vapor stream leaving the reboiler is impractical. Instead, the controls must influence the boilup via the heat input to the reboiler. In Figure 1.9, the heat is supplied by steam, and a control valve is provided on the steam supply.

Bottoms reflux L_B (an input term). This is the liquid flow leaving the lower separation section within the column. The controls at the bottom of the column have no way to influence the bottoms liquid L_B.

The unsteady-state material balance around the reboiler is written as follows:

$$L_B - (B + V) = \frac{dH_B}{dt},$$

where H_B is the bottoms holdup (mole).

1.5.2. Control Configurations

The two manipulated variables, the bottoms flow B and the boilup V, associated with the reboiler are used to control the following two variables:

Bottoms composition. When the bottoms product is a salable product, good bottoms composition control is crucial.

Bottoms level. Rarely does the bottoms level affect any term in the profit-and-loss statement.

In selecting the control configuration, controlling the bottoms composition must take priority, as reflected in the following approach:

1. Determine if the bottoms composition is to be controlled by manipulating the boilup V or by manipulating the bottoms flow B. This takes precedence over the usual preference to control level by manipulating the larger of the two flows (B or V).

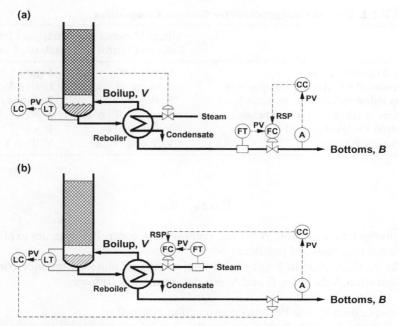

Figure 1.10. Control configurations for bottoms composition. (a) Direct material balance control. (b) Indirect material balance control.

2. Control bottoms level with the other flow. However, level cannot be controlled by manipulating a very small flow. If $V/B \ll 1$, drum level cannot be controlled by manipulating V. If $V/B \gg 1$, drum level cannot be controlled by manipulating B.

Figure 1.10 presents the two possible control configurations, which are designated *direct material balance control* and *indirect material balance control*. The bottoms flow B appears explicitly in the total material balance for the column:

$$F = D + B.$$

The terms direct material balance control and indirect material balance control pertain to how the value of the bottoms flow is obtained. Table 1.2 summarizes the attributes of the two configurations.

The configuration in Figure 1.10a is the direct material balance control configuration. Values for B and V are determined as follows:

B—specified by the bottoms composition controller;

V—determined by the level controller to satisfy the steady-state material balance for the reboiler:

TABLE 1.2. Control Configurations for Bottoms Composition

	Direct Material Balance Control	Indirect Material Balance Control
Control configuration	Figure 1.10a	Figure 1.10b
Manipulated variable for composition	Bottoms B	Boilup V
Manipulated variable for bottoms level	Boilup V	Bottoms B
Solution of reboiler material balance	$V = L_B - B$	$B = L_B - V$
Preferred for level control if	$V > B$	$B > V$
Impractical if	$V/B \ll 1$	$V/B \gg 1$

$$V = L_B - B.$$

The manipulated variable B for the composition controller appears explicitly in the column material balance.

The configuration in Figure 1.10b is the indirect material balance control configuration. Values for B and V are determined as follows:

V—specified by the bottoms composition controller;

B—determined by the level controller to satisfy the steady-state material balance for the reboiler:

$$B = L_B - V.$$

The manipulated variable V for the composition controller does not appear explicitly in the column material balance. Instead, the composition controller specifies V, from which the level controller determines the value of B.

1.6. COMPONENT MATERIAL BALANCES

Herein component material balances will only be developed for the entire column. Component material balances can be made for the condenser and the reboiler, but these seem to have no significant implications for control.

1.6.1. Steady-State Equations

A component material balance can be written for each component in the feed. For binary distillation, there are two components (light and heavy), hence two equations:

$$\text{Light component: } F z_L = D y_L + B x_L,$$

$$\text{Heavy component: } F z_H = D y_H + B x_H.$$

The respective mole fractions must sum to unity:

$$x_L + x_H = 1,$$
$$y_L + y_H = 1,$$
$$z_L + z_H = 1.$$

Summing the above two component material balance equations gives the total material balance:

$$F(z_L + z_H) = D(y_L + y_H) + B(x_L + x_H),$$
$$F = D + B.$$

To obtain a set of independent equations, the total material balance can be used in lieu of either of the component material balances.

1.6.2. Degrees of Freedom

The analysis will be based on the following two independent equations:

$$\text{Total material balance}: F = D + B,$$
$$\text{Component material balance, light } F\, z_L = D\, y_L + B\, x_L.$$

A fixed service is assumed, which means that the feed flow F and the feed composition z_L are otherwise specified. The degrees of freedom are as follows:

Number of variables: 4 ($B, D, y_L,$ and x_L)
Number of equations: 2
Degrees of freedom: $4 - 2 = 2$

For control, this means that independent targets can be provided for two of the four variables ($B, D, y_L,$ and x_L). However, this does not mean "any two."

1.6.3. Control Options

For a total of four variables, there are six possible subsets of two. But for the distillation column, it is possible to provide independent targets for only five of the six possible subsets:

Subset 1: D and y_L
Subset 2: D and x_L
Subset 3: B and y_L
Subset 4: B and x_L
Subset 5: y_L and x_L

Because the mole fractions must sum to unity, y_H can be used in lieu of y_L and/or x_H in lieu of x_L.

The sixth possible subset of two is D and B. However, degrees of freedom also apply to subsets of the equations. One of the equations in the set is the total material balance. This equation does not permit targets for D and B to be specified independently.

1.6.4. Composition Control

The degrees of freedom analysis suggests that the following are possible:

1. For one of the product streams, specify a target for the flow and a target for the composition:
   - Specify distillate flow D and distillate composition y_L or y_H.
   - Specify bottoms flow B and bottoms composition x_L or x_H.
   In practice, this is not common.
2. Specify a target for the flow of either product stream and a target for the composition of the other product stream:
   - Specify distillate flow D and bottoms composition x_L or x_H.
   - Specify bottoms flow B and distillate composition y_L or y_H.
   This is commonly used for single-end composition control.
3. For both product streams, specify a target for the composition.
   - Specify distillate composition y_L or y_H and bottoms composition x_L or x_H.
   This is double-end composition control.

The latter combination is of particular interest. Specifically, the degrees of freedom are sufficient to control both compositions.

1.6.5. Double-End Composition Control

Many difficulties were experienced in the early attempts, and applications of double-end composition control remained rare until the 1970s. The degrees of freedom analysis only suggests that something is possible; it does not propose a control configuration that will be successful.

The root of most problems was interaction between the two composition loops. There is inherently some interaction in every double-end composition control configuration. Any change that affects the composition of one product stream will have some effect on the composition of the other product stream. For each component of the feed, if one additional unit of that component is removed in the distillate stream, then one unit less of that component must be removed in the bottoms stream.

The degree of interaction depends on many factors, including the purities of the products, the external reflux ratio, and the relative volatility of the components. A proposed control configuration must be analyzed in light of the degree of interaction exhibited by the column on which it will be installed. Eventually, double-end composition control will be implemented on about 80% of the distillation columns.

1.6.6. Values for Targets

Suppose the degrees of freedom analysis suggests that two targets can be independently specified. This does not mean that all combinations of values for the targets are acceptable.

Probably the best way to express this is that the values for the targets must be "within reason." Basically, this means that the values specified for the targets do not result in values for other variables that are impossible to attain. For distillation applications, the values specified for the targets must not give results such as the following:

1. A value for a composition that is less than 0% or greater than 100%.
2. A value for a flow that is negative. Reversible flow is not permitted for the distillate product, the bottoms product, reflux, and so on.

Mathematically, negative values could certainly be computed. In the formulation of the problem, inequalities such as $D \geq 0$, $B \geq 0$, and $0 \leq y_L \leq 1$ should be included. But instead of writing these explicitly, phrases such as "within reason" are sometimes applied.

1.6.7. Recovery

The recovery is the fraction of the feed that goes to a respective product stream. For the distillate product, the recovery is D/F; for the bottoms product, the recovery is B/F. The recovery is often an important measure of column efficiency. If the distillate product is the salable product, improvements in the distillate recovery increase the amount of the desirable product that is available for sale.

The recovery is related to the various compositions (feed, distillate, and bottoms) and vice versa. This is vividly illustrated when the component material balance for the light component is rewritten as follows:

$$F\, z_L = D\, y_L + B\, x_L = D\, y_L + (F - D)x_L = D(y_L - x_L) + F\, x_L$$

$$F(z_L - x_L) = D(y_L - x_L)$$

$$\frac{D}{F} = \frac{z_L - x_L}{y_L - x_L}.$$

When controlling product compositions, the usual approach is to focus on the energy terms (reflux and boilup). However, ignoring the role of the column material balance is an invitation for problems.

1.7. ENERGY AND THE SEPARATION FACTOR

In a distillation, column separation is attained by successive stages that essentially involve vaporization of a liquid and condensation of a vapor. Both involve energy. Except in towers with side heaters and/or side coolers, the energy for vaporization is provided largely by the reboiler, and the energy released by condensation is removed largely by the condenser.

Since energy is providing separation, the intuitive conclusion is that product compositions must be controlled through energy, which in most towers means the boilup and the reflux. The result is the double-end composition control configuration in Figure 1.11, in which the distillate composition is controlled by adjusting the reflux and the bottoms composition is controlled by adjusting the boilup. This is indirect material balance control for both product compositions—both the distillate flow and the bottoms flow are determined by the difference in two energy terms.

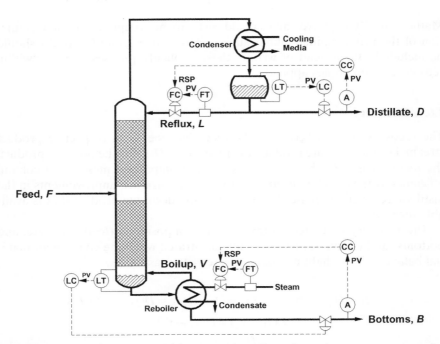

Figure 1.11. Double-end composition control configuration using an energy term for each product composition.

In a subsequent chapter, interaction analysis will be introduced as the tool for analyzing the degree of interaction in a proposed control configuration for distillation. In most cases, the degree of interaction for the configuration in Figure 1.11 is high, which translates into operational problems in the field. The degree of interaction is usually much lower for configurations in which one composition is controlled by manipulating an energy term and the other composition is controlled by manipulating a product draw.

Sometimes, it is difficult to convince people that what seems intuitive is perhaps off-base, at least in some cases. Distillation is a complex process, which complicates making the argument that controlling a product composition with a product draw is not only possible but appropriate. For double-end composition control, one of the compositions must be controlled by an energy term (D and B are not independent variables). But the other composition can be controlled using a product draw, and in most towers, this provides the least degree of interaction.

The objective of this section is to present the argument that controlling a product composition with a product draw just might make sense. To make this argument, a relationship between separation and energy is required. This is a complex relationship, even for binary distillation. The objective herein is to provide an insight into the issues, not to use the relationship for computational purposes. To keep it simple, the presentation will rely on the following:

1. An approximate relationship between separation and energy;
2. A binary separation.

However, the conclusions apply to multicomponent columns as well.

1.7.1. Fenske Equation

Most approximate relationships for separation are derived in some manner from the Fenske equation that relates the product compositions to the relative volatility and the number of stages:

$$\frac{y_L(1-x_L)}{x_L(1-y_L)} = \frac{y_L \ x_H}{x_L \ y_H} = \frac{y_L/x_L}{y_H/x_H} = \alpha^n,$$

where

n = number of theoretical stages;

α = relative volatility (ratio of vapor pressures) of the light component relative to the heavy component.

Unfortunately, the Fenske equation has a serious restriction—it only applies at total reflux.

1.7.2. Separation Factor

When the tower is not operating on total reflux, the term α^n in the Fenske equation is replaced by the separation factor S:

$$S = \frac{y_L(1-x_L)}{x_L(1-y_L)} = \frac{y_L \, x_H}{x_L \, y_H} = \frac{y_L/x_L}{y_H/x_H}.$$

For most columns, the numerical value of the separation factor will be large, especially if the products are low in impurities (y_H in the distillate product; x_L in the bottoms product). Suppose both products are 95% pure, which is not an especially high purity. The value of the separation factor is

$$y_L = 0.95,$$
$$y_H = 0.05,$$
$$x_L = 0.05,$$
$$x_H = 0.95,$$
$$S = \frac{y_L/x_L}{y_H/x_H} = \frac{0.95/0.05}{0.05/0.95} = 361.$$

In practice, values of 1000 or more for the separation factor are typical.

1.7.3. Separation Factor and Control

The above example computed the separation factor from the distillate and bottoms compositions. But in practice, the distillate and bottoms compositions depend on the separation factor and the column material balances.

The value of the separation factor depends on the following:

Number of theoretical stages n. Largely determined by the column design; operating variables have only a minor influence.

Relative volatility α. Depends primarily on the materials being separated. Column pressure has some influence and is occasionally used for optimization but never for regulatory control.

Energy input Q. Variable that the control system can influence through the reflux and boilup rates.

In order to affect the separation factor in an operating tower, the control system must change the energy terms. In a sense, this reinforces one's intuition that product compositions should be controlled through energy.

Although a few relationships have been proposed, relating the separation factor to the number of theoretical stages n, the relative volatility α, and the energy (either as reflux ratio or boilup ratio) is a challenge. Fortunately, this is not necessary for the discussion that follows—again, the objective is to gain insight, not to perform computations.

1.7.4. Coupling Material Balance with Separation

For a binary tower, the following equations relate the product compositions (y_L and x_L) to the D/F ratio (the recovery for the distillate product) and the separation factor S:

$$\text{Material balance: } \frac{D}{F} = \frac{z_L - x_L}{y_L - x_L}$$

$$\text{Separation: } S = \frac{y_L(1 - x_L)}{x_L(1 - y_L)}$$

With four unknowns (D, S, y_L, and x_L) in two equations, the solution can be viewed in two ways:

1. Though its final control elements, the control system specifies the product draws (which determine D/F) and the energy terms (which determine the separation factor S). The above two equations can be solved for the product compositions y_L and x_L.
2. In a double-end composition control application, the product specifications provide targets for y_L and x_L. The above two equations can be solved for the recovery D/F and the separation factor S. Basically, this is the solution that the controls must obtain in basically a trial-and-error fashion.

Even for binary columns, the solution of the two equations requires iterative procedures. Consequently, these equations are of little (or no) computational value. However, they provide the basis for gaining insight into the control options for a column.

1.7.5. Approximations in Separation Factor Equation

In many columns, the impurities y_H and x_L in both products are small, which permits the following approximations to be made:

$$1 - y_H \cong 1,$$
$$1 - x_L \cong 1.$$

With these approximations, the expression for the separation factor simplifies to the following:

$$S = \frac{y_L / x_L}{y_H / x_H} = \frac{(1 - y_H)/x_L}{y_H /(1 - x_L)} \cong \frac{1}{y_H \, x_L}.$$

In a previous example, the purity of both product streams was 95%, giving a separation factor of 361. With the above approximation, the separation factor is

$$S = \frac{1}{0.05 \times 0.05} = 400.$$

The higher the purity, the less the difference.

1.7.6. Logarithmic Equation for Separation Factor

When analyzing the expressions for the separation factor, the nonlinear nature of the equation leads to complications. But when the impurities in both products are small, expressing the relationship in terms of logarithms gives a linear result:

$$\ln S = -\ln y_H - \ln x_L = (-\ln y_H) + (-\ln x_L).$$

The compositions y_H and x_L are both less than 1, so the quantities $(-\ln y_H)$ and $(-\ln x_L)$ are positive values.

1.7.7. Graphical Representation

The objective of the graphical representation in Figure 1.12 is to illustrate this point. There are two scales:

Upper scale. The composition y_H of the impurity in the distillate.
Lower scale. The composition x_L of the impurity in the bottoms.

Both scales are logarithmic. The two scales are joined for a composition of 1.0 (which is zero on a log scale).

Starting with values for the separation factor S and the distillate draw D gives product compositions of y_H impurity in the distillate and x_L impurity in the bottoms. These are represented on the graph as the distances $(-\ln y_H)$ for the impurity in the distillate and $(-\ln x_L)$ for the impurity in the bottoms. The sum of these two distances is $(\ln S)$. Starting from the solution designated as the "base case" in Figure 1.12, the effect of increasing the separation factor S and then increasing the distillate flow D will be illustrated.

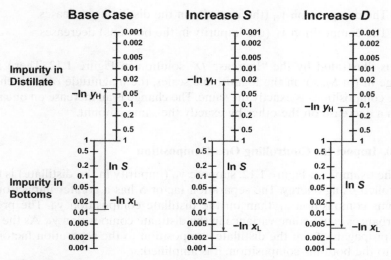

Figure 1.12. Effect of the separation factor S and the distillate draw D on product compositions.

1.7.8. Increasing the Separation Factor

The impact of increasing the separation factor is simple: the value of $(\ln S)$ increases. This means that either $(-\ln y_H)$ increases, $(-\ln x_L)$ increases, or both. In practice, there is some increase in both, as illustrated by the "Increase S" solution in Figure 1.12.

However, the increase is usually not by the same amount. The extremes for the possibilities are as follows:

- The major impact is on y_H (the impurity in the distillate), with little change in x_L (the impurity in the bottoms).
- The major impact is on x_L (the impurity in the bottoms), with little change in y_H (the impurity in the distillate).

However, it is also possible for the impact to be about evenly distributed between y_H and x_L. No general statements can be made about what result to expect. The only way to obtain answers is to use a distillation column model to examine the effect of increasing the energy input to the column.

1.7.9. Increasing the Distillate Draw

If the separation factor S is held constant, increasing the distillate draw D increases the concentration of the heavy component in every separation stage. Consequently, the results will be as follows:

1. The composition y_H (the impurity in the distillate) increases.
2. The composition x_L (the impurity in the bottoms) decreases.

This is illustrated by the "Increase D" solution in Figure 1.12. There is no change in ($\ln S$), so on the logarithmic scales, the magnitude of each change in the compositions is exactly the same. The change is an increase on one scale, but is a decrease on the other by exactly the same amount.

1.7.10. Impact for Controlling One Composition

For the example in Figure 1.12, suppose y_H (impurity in the distillate) is to be controlled using energy. The separation factor S has a greater impact on the bottoms composition x_L than on the distillate composition y_H. The process disturbances cause some variance in the distillate composition y_H. As the variance propagates from the distillate composition to the separation factor and then to the bottoms composition, it is amplified:

1. To compensate for the variance in y_H, the composition controller changes the separation factor S (through changing the energy input to the column). The smaller the effect of the separation factor S on a composition, the larger the changes required in the separation factor S in order to maintain the composition at its target.
2. The changes in separation factor S will affect the bottoms composition. For the example, in Figure 1.12, the effect of the separation factor S on x_L is larger than its effect on y_H. This significantly increases the variance in x_L.

Since only y_H is being controlled, is variance in x_L of any concern? Variance in x_L is likely to impact some downstream operation. Situations where reducing the variance in one variable greatly amplifies the variance in another should be avoided.

1.7.11. Double-End Composition Control

Controlling only one composition can usually be accomplished via the energy streams that affect the separation factor S. Potentially, the control actions taken to control that composition could propagate significant variance to the other composition. But since this composition is not being controlled, there will be no closed-loop response to the variations propagated to the other stream.

When both compositions are to be controlled, the issues pertaining to interaction must be resolved. Normally, the composition most affected by the separation factor must be controlled by making changes in the energy streams. For the case illustrated in Figure 1.12, this means that the bottoms composition must be controlled using the energy streams. Changes in the separation factor S have more influence on the bottoms composition than on the distillate composition.

If the bottoms composition is controlled via the separation factor, how does one control the overhead composition? The relationships on which the graphs in Figure 1.12 are based suggest that the distillate composition can only be controlled by changing a product draw. In essence, the bottoms composition is controlled through changes in energy; the distillate composition is controlled through the material balance.

1.8. MULTICOMPONENT DISTILLATION

One always likes to start with the simple, which in distillation means binary distillation. A few binary distillation columns are found in production facilities, but most are multicomponent.

The next section will discuss the stage-by-stage separation models that are now routinely used in column design. These provide very accurate solutions, but at the expense of considerable complexity. Design and control are fundamentally different. For a column in a specified service (feed flow and composition), the problems are stated as follows:

Design. Calculate the reflux, boilup, and so on, required to give specified product compositions.

Control. The current operating conditions in the tower are known (reflux flow, boilup, product compositions, etc.). Calculate the change in the manipulated variable, such as the boilup, required to change the controlled variable, such as the bottoms composition, from its current value to its target.

To summarize, design works on actual values, and for this, accuracy is crucial. However, control works on changes (a change in the manipulated variable leads to a change in the controlled variable), and especially when the changes are small, approximations would certainly be acceptable.

The Hengstebeck approximation is one example that will be explained shortly. Prior to the computer era, columns were designed based on such approximations, but detailed models are now used in lieu of such approximations.

1.8.1. Heavy and Light Keys

In binary distillation, the components are referred to as the light component and the heavy component. The objective is to separate these two components. In multicomponent distillation, the corresponding terms are "light key" and "heavy key." A column effects a separation between the two keys.

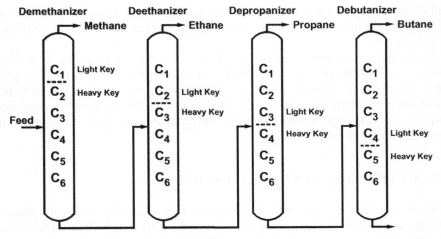

Figure 1.13. Columns in series.

Figure 1.13 illustrates a sequence of columns such as found in a gas plant. The feed to the first column (the demethanizer) is a mixture of methane (C_1), ethane (C_2), propane (C_3), butane (C_4), and so on. The columns and their key components are as follows:

Demethanizer. Separates methane (the light key) from ethane (the heavy key).

Deethanizer. Separates ethane (the light key) from propane (the heavy key).

Depropanizer. Separates propane (the light key) from butane (the heavy key).

Debutanizer. Separates butane (the light key) from pentane (the heavy key).

In binary distillation, the light and heavy components appear in both product streams. A component that appears in both product streams is said to be a "distributed component." In the Hengstebeck approximation, only the light and heavy keys are distributed, the assumptions being as follows:

1. All components of the feed that are lighter than the light key leave with the distillate product. Basically, these components are treated as noncondensible gases.
2. All components of the feed that are heavier than the heavy key leave with the bottoms product. Basically, these components are treated as nonvolatile liquids.

Consider the depropanizer in the separation train in Figure 1.13. The light key is propane (C_3); the heavy key is butane (C_4). All methane and ethane leave with the distillate product; all pentane and heavier components leave with the bottoms.

1.8.2. Components

When using the Hengstebeck approximation for the depropanizer in Figure 1.13, the four components are as follows:

Lighter-than-light key (LL). This includes all methane and all ethane. These components leave entirely with the distillate product. This is a pseudo-component whose composition in the distillate product is y_{LL}. None of these components appear in the bottoms, so x_{LL} is zero.

Light key (L). This is propane. This component appears in both the distillate product and the bottoms product, and thus is a distributed component. In a multicomponent system, y_L is the composition of the light key in the distillate and x_L is the composition of the light key in the bottoms.

Heavy key (H). This is butane. This component appears in both the distillate product and the bottoms product and thus is a distributed component. In a multicomponent system, y_H is the composition of the heavy key in the distillate and x_H is the composition of the heavy key in the bottoms.

Heavier-than-heavy key (HH). This includes all pentane and heavier components. These components leave entirely with the bottoms product. This is a pseudocomponent whose composition in the bottoms product is x_{HH}. None of these components appear in the distillate, so y_{HH} is zero.

1.8.3. Component Material Balances

The light key (L) and the heavy key (H) are real components; the lighter-than-light (LL) and the heavier-than-heavy (HH) are pseudocomponents. A component material balance can be written for each:

$$F\, z_{LL} = D\, y_{LL},$$
$$F\, z_L = D\, y_L + B\, x_L,$$
$$F\, z_H = D\, y_H + B\, x_H,$$
$$F\, z_{HH} = B\, x_{HH}.$$

The equations for the light key and the heavy key are identical to those written for the light and heavy components of binary distillation.

1.8.4. Separation Factor

The objective of the Hengstebeck approximation is to permit the relationships developed for binary distillation to be applied to multicomponent distillation. For example, the equation for the separation factor is still

$$S = \frac{y_L}{x_L} \frac{x_H}{y_H} = \frac{y_L / x_L}{y_H / x_H}.$$

For multicomponent separations, y_L and x_L pertain to the light key; y_H and x_H pertain to the heavy key.

1.9. STAGE-BY-STAGE SEPARATION MODEL

The reality is that all components of the feed to a tower appear to some extent in both product streams. For a depropanizer, the amount of ethane in the bottoms will be extremely small, but some will be present. The amount of methane in the bottoms will be even smaller, but some will be present. Similar statements can be made with regard to the pentane in the distillate stream. Assuming the composition of these minor components to be zero is not always acceptable.

Another issue arises when isomers are present. Consider butane. In most gas plants, butane is primarily n-butane. However, some isobutane is present. Isobutane is more volatile (has a lower boiling point) than n-butane. Consequently, the ratio of isobutane to n-butane in the distillate will be higher than their ratio in the feed.

1.9.1. Separation Model

The stage-by-stage separation model is based on the following equations:

1. A component material balance is written for each component on each stage. If the feed to the column contains 10 components and there are 30 stages in the column, this gives 300 equations.
2. An energy balance is written for each stage (equimolal overflow is not assumed).
3. Realistic vapor–liquid equilibrium relationships can be used. Without such relationships, the relative volatility is assumed to be constant, which is rarely the case.

This gives a large number of nonlinear equations. Only computers can solve such equations. Today, commercial software packages are available that are specifically designed to solve the equations that arise in distillation, and most companies have standardized on one (or perhaps two) of these.

1.9.2. Issues for Control

The stage-by-stage separation models are occasionally used in on-line optimization and similar undertakings. Incorporating into regulatory control configurations poses two problems:

High dimensionality. The total number of equations is very large. For 10 components and 30 stages, the number of equations is in excess of 300. The concepts for controlling multivariable processes are well known, but high dimensionalities present a variety of problems, including numerical difficulties.

Nonlinear equations. The vapor–liquid equilibrium relationships are highly nonlinear. Unfortunately, most of the currently available control technologies are based on linear systems theory.

Difficulties such as these can certainly be overcome. However, there must be an incentive to do so. The improved accuracy of the stage-by-stage separation models led to improved column designs, which provided the incentives to develop techniques specifically for solving the model equations. But regulatory control depends primarily on repeatability, not accuracy. To date, no incentives have been identified that justify developing methods to incorporate the stage-by-stage separation models into regulatory control configurations.

1.9.3. Start with a Column Model

Even though the stage-by-stage separation model will not be used directly in the regulatory control configuration, developing such a model must be the starting point for any control effort directed to a distillation column.

When analyzing a control problem associated with a distillation column, the first step is to make sure the column is capable of doing what is desired. A common practice in production facilities is to blame all problems on the control system. Process problems often lead to the control system being unable to maintain a process variable at its target. But if the process is unable to attain the target value, control efforts directed at the problem are doomed to failure. Distillation is a complex unit operation that offers many possibilities for problems to arise. Some of these problems can be very subtle, and some problems will only arise under certain situations.

Normally, the data set for a stage-by-stage separation model is part of the "deliverables" from the design team. Ideally, the startup effort should include collecting data from the column and calibrating the model to the process, but this is not always the case. Starting with whatever is available, one proceeds as follows:

1. Collect current operational data from the tower (flows, temperatures, compositions, etc.).

2. Calibrate the separation model to the process by adjusting parameters such as stage efficiencies.

If the column performance is far different from what the model suggests, this must be resolved before proceeding with any true control work.

1.9.4. Steady-State Issues in Control

The common impression is that regulatory control is only concerned with process dynamics. This view is reinforced by the typical academic course on "process control," which is in reality a mathematics course on linear systems theory. But in most applications, the key issues pertain to the steady-state behavior of the process, not its dynamics. Consequently, much that is relevant to regulatory control of a distillation column can be understood from its steady-state model.

One relevant characteristic is the sensitivity. If you increase the energy flows (energy in at the reboiler and energy out at the condenser), you would expect the impurities in both product streams to decrease. But will the major effect be in the distillate composition, will it be in the bottoms composition, or will the impact be about the same in both product compositions? Distillation is a complex unit operation, so questions such as these can only be answered with confidence when the answers are obtained via a good separation model. This is especially true in complex towers and towers separating nonideal mixtures.

In distillation columns, some degree of interaction always exists between the product compositions. When double-end composition control is being attempted, this interaction must be analyzed very carefully. There are two aspects of interaction—steady state and dynamic. The dynamics of the composition loops will be about the same, which makes the steady-state aspects very significant. The analysis of this interaction can be based entirely on results obtained from the separation model.

1.9.5. Limitations

Engineering involves obtaining numerical answers to numerical problems. In this regard, the stage-by-stage separation models are superb. Probably, the main concern is the quality of some of the relationships (vapor–liquid equilibrium data, heat capacity equations, etc.). Even minor changes in these relationships can give significantly different results.

But suppose one's objective is to obtain insight into how a specific tower behaves, or possibly to improve one's understanding of distillation in general? The stage-by-stage separation models are not very useful. One can obtain a series of solutions by changing certain parameters and examining their effect on the results. One is quickly inundated with data. The stage-by-stage separation models are very good at one thing—determining the numerical solution to the column equations for a specific situation (feed rate, reflux ratio, etc.).

1.9.6. Depropanizer Model

In order to illustrate how information relevant to control can be obtained from the steady-state model of a distillation column, a simplified version of a depropanizer from a production facility will be used. Just to avoid carrying too many numbers, the following simplifications are made:

1. The feed (from a deethanizer) contained very little ethane and almost no methane. The methane composition is set to zero.
2. All components heavier than pentane are treated as pentane.

The feed contains only four components, their compositions being the following:

Ethane (C_2): 0.4 mol%
Propane (C_3): 23.0 mol% (the light key)
Butane (C_4): 37.0 mol% (the heavy key)
Pentane (C_5): 39.6 mol%

1.9.7. Separation Sections

The upper separation section (above the feed stage) has 11 ideal stages; the lower separation section has 9. The column has a total condenser (distillate product is liquid). Including the reboiler, the column has a total of 21 ideal stages. Stages will be numbered from the top of the tower. That is, stage 1 is at the top of the upper separation section; stage 20 is at the bottom of the lower separation section; stage 21 is the reboiler.

The column pressure is 16.0 barg. The overhead product is primarily propane, so the overhead temperature will be approximately the boiling point of propane (50.3°C at 16.0 barg). Coefficients for the relationships for vapor pressures, heat capacities, and so on, are obtained from Yaws [1].

In distillation calculations, a common approach is to base the calculations on a feed rate of 100 mol per unit time (hour, minute, etc). Herein mol/h will be used. The feed enters as a liquid under pressure at 105°C. About 10% of the feed flashes upon entry into the tower.

1.9.8. Base Case

One of the first steps in the analysis of any control problem with a distillation column is to develop a column simulation that matches the current plant

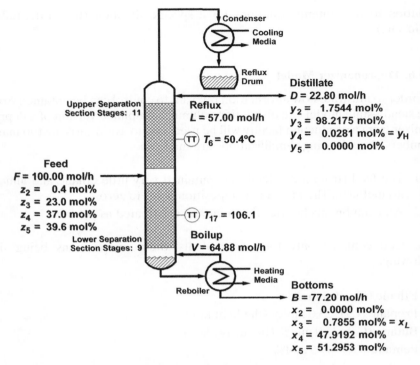

Figure 1.14. Depropanizer model base case solution.

operating conditions. This solution becomes the "base case" for subsequent analyses. For the depropanizer, the solution for the base case is computed for the following conditions:

$$\text{Distillate flow } D = 22.80 \text{ mol/h}$$

$$\text{External reflux ratio } (L/D) = 2.5 \text{ (since } D = 22.80, L = 57.00 \text{ mol/h).}$$

Figure 1.14 summarizes the solution of the steady-state model for the base case.

A subsequent discussion on product compositions will explain why the column is operated in this manner, but briefly, the objective is for the propane product to contain as much ethane as the specifications permit (and consequently very little butane) and for the butane product (to the next column) to contain as much propane as the specifications permit.

In practice, one cannot assume that the column is well-designed for its current service. Although design mistakes are occasionally made, the most likely explanation is that the column is not being operated for the service for which it was designed.

1.9.9. Utility in Control Analyses

When any operational problem arises for a distillation column, the basic controls are frequently viewed as the culprit. Sometimes they are, but not always. The analysis of the problem should begin with a stage-by-stage column simulation. Especially when the column is not operating in the service for which it was designed, two aspects must be verified:

1. The column is performing in a manner consistent with its design.
2. The column can deliver the performance being demanded by current operations.

Beyond these, there are other uses of the stage-by-stage simulation, including the following:

Temperature profile. Improper location of the control stages leads to problems with temperature controls.

Internal flows. The internal vapor and liquid flows must be within the limits imposed by the tower internals.

Sensitivities. Distillation is a complex unit operation with significant interaction between the operating variables.

1.9.10. Temperature Profile

A common practice is to use temperature measurements in either or both of the following manners:

Upper control stage. The temperature of this stage (from the upper separation section) is used as the measured variable for a temperature controller that adjusts either the reflux flow L or the distillate flow D.

Lower control stage. The temperature of this stage (from the lower separation section) is used as the measured variable for a temperature controller that adjusts either the heat input to the reboiler (and consequently the boilup flow V) or the bottoms flow B.

When either or both of these are used, the temperature profile within the tower must be examined. The temperature on each stage is computed through the energy balance that is incorporated into the stage-by-stage calculations. The temperature profile is obtained by plotting these temperatures, the result being the graph presented in Figure 1.15.

The graph also indicates the location of the control stages. The issues pertaining to using stage temperatures in control configurations will be examined in the next chapter.

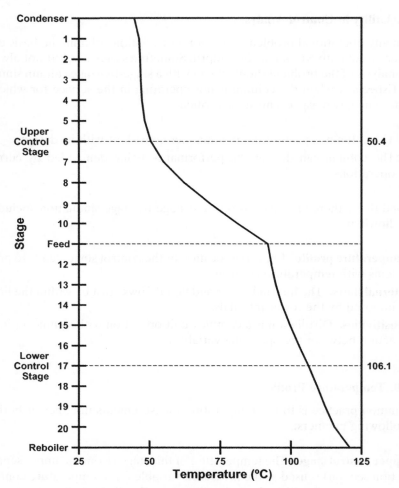

Figure 1.15. Temperature profile.

1.9.11. Internal Vapor and Liquid Flows

The stage-by-stage calculations also provide values for the vapor and liquid flows leaving each stage. Figure 1.16 presents the vapor and liquid flows for the depropanizer plotted in a manner similar to the stage temperatures. The most noticeable change is at the feed stage:

   Liquid flow. About 90% of the feed contributes to the liquid flow, so the liquid flow below the feed stage is significantly higher than the liquid flow above the feed stage.

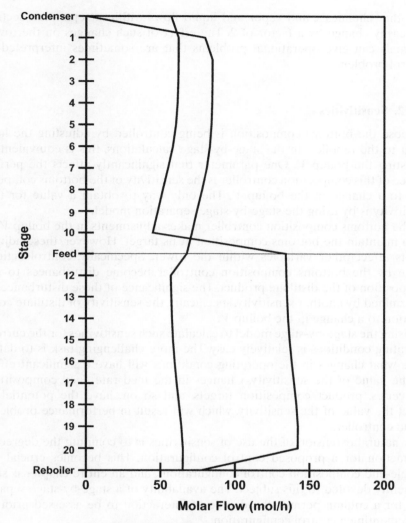

Figure 1.16. Vapor and liquid flows.

Vapor flow. About 10% of the feed contributes to the vapor flow, so the vapor flow above the feed stage is only slightly higher than the vapor flow below the feed stage.

Although not constant, the changes in liquid and vapor flows within each separation section are nominal. This is generally the case for mixtures that are close to ideal. Mixtures of hydrocarbons such as propane and butane deviate only slightly from ideal. Larger departures are normally the case for other chemicals, and vapor and liquid flows within a separation section can easily change by a factor of 2. The effect of such changes on the tower internals can give operational problems that are sometimes interpreted as control problems.

1.9.12. Sensitivities

Suppose the bottoms composition is being controlled by adjusting the heat input to the reboiler. In the stage-by-stage calculations, this is equivalent to adjusting the boilup V. One parameter that significantly affects the performance of this composition controller is the sensitivity of the bottoms composition to a change in the boilup V. The only way to obtain a value for this sensitivity is by using the stage-by-stage separation model.

The bottoms composition controller makes adjustments in the boilup V so as to maintain the bottoms composition at its target. However, these adjustments affect other variables within the tower. Specifically, control actions taken by the bottoms composition controller become disturbances to the composition of the distillate product. The significance of these disturbances is determined by another sensitivity, specifically, the sensitivity of distillate composition to a change in the boilup V.

Using the stage-by-stage model to calculate such sensitivities for the current operating conditions is relatively easy. The more challenging task is to determine what changes in the operating conditions will have a significant effect on the value of the sensitivity. Changes in the feed rate, feed composition, recoveries, product composition targets, and so on, have the potential to affect the value of the sensitivity, which will result in performance problems in the controller.

A natural extension of the use of sensitivities is to compute the degree of interaction for a proposed control configuration. This becomes crucial for double-end composition control configurations, and an entire chapter is subsequently devoted to this subject. The availability of a stage-by-stage separation for a column permits the degree of interaction to be assessed prior to implementing a control configuration.

1.9.13. Precision

In this context, precision will be used as in C++—the number of digits after the decimal point for representing numerical values. Sensitivities are computed from the difference in two values. Distillation is nonlinear, so the difference in the two values must be the result of a small difference in variables such as boilup, reflux, and product flow. Consequently, the difference in the two values will be small.

Herein the values of flows, compositions, temperatures, and so on, will be routinely represented to a greater precision than justified by the separation model. Computing sensitivities depends on the ability of the model to translate a small change in one variable to a small change in another. Models can generally do this better than the accuracy of the individual values. This is much like the accuracy versus repeatability of a measurement device—the repeatability is usually better than the accuracy.

1.10. FORMULATION OF THE CONTROL PROBLEM

Distillation columns are relatively complex unit operations with a large number of permutations. At this point, the column in Figure 1.17 is arbitrarily used as the starting point. The key aspects of the configuration in Figure 1.17 are as follows:

- The tower is a two-product tower.
- Both product streams are liquid (condenser is a total condenser).
- The reflux drum is partially filled (the reflux drum level must be measured and controlled).
- The condenser transfers heat to cooling water.
- The reboiler is heated with steam.

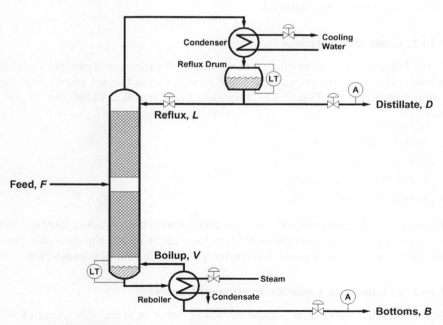

Figure 1.17. Controlled and manipulated variables for a two-product tower.

1.10.1. Reboiler and Condenser

Using a steam-heated reboiler in the P&I diagram in Figure 1.17 is reasonable—steam is the most common heating medium used in production facilities. Alternatives such as hot oil and fired heaters will be discussed along with various reboiler configurations in the subsequent chapter devoted to reboilers.

There are a couple of issues pertaining to the condenser arrangement in Figure 1.17:

- Although water-cooled condensers are probably installed most frequently, air-cooled condensers are common.
- Varying the water flow through the condenser raises lots of issues. Alternatives such as hot gas bypass arrangements are often installed, especially when natural water is used as the cooling media.

All of these are discussed in the subsequent chapter on pressure control and condensers. Some mechanism by which the control system can vary the heat transfer rate in the condenser is required, but the exact nature of that mechanism has little impact on the remaining control issues for the tower.

The "default" condenser arrangement used in most illustrations within this book is a water-cooled condenser with a control valve on the cooling water, as in Figure 1.17. The reason: this is the simplest to draw. Another simple arrangement is a control valve in the overhead vapor line, but this arrangement is not commonly installed.

1.10.2. Controlled Variables

In the language of control engineers, a controlled variable is a process variable whose value is to be maintained at or near a target (or set point). For the column illustrated in Figure 1.17, there are five controlled variables:

1. bottoms level,
2. reflux drum level,
3. column pressure,
4. distillate composition,
5. bottoms composition.

Figure 1.17 indicates composition measurements on both product streams. But as noted previously, practical considerations often locate the analyzer elsewhere or even utilize a control stage temperature in lieu of composition.

1.10.3. Manipulated Variables: Instrument Context

A manipulated variable is a variable whose value is at the discretion of the control system. At the hardware level, these are the physical outputs of the controls, which are signals that drive a final control element. In distillation columns, the final control element is usually a valve, but occasionally is a variable speed drive, a power regulator to an electric heater, and so on.

At the instrument level, the five manipulated variables for the distillation column illustrated in Figure 1.17 are as follows:

1. distillate valve opening,
2. bottoms valve opening,
3. reflux valve opening,
4. condenser cooling water valve opening,
5. reboiler steam valve opening.

1.10.4. Manipulated Variables: Process Context

For each manipulated variable, there is a process variable that corresponds to the instrument variable. Consider the stage-by-stage separation models. One never specifies the distillate valve opening; one specifies the distillate flow. One could always consider the process variable to be the flow through the control valve, but this is not very satisfactory for the condenser or the reboiler. Instead, the choices are the following:

Condenser. The cooling water flow affects the condenser heat transfer rate Q_C, which in turn affects the condensation rate within the condenser. At steady state, the overhead vapor rate V_C must be consistent with the condensation rate. Herein, the overhead vapor rate V_C will generally be used as the manipulated variable associated with the condenser.

Reboiler. The steam flow determines the reboiler heat transfer rate Q_R, which in turn affects the vaporization rate within the reboiler. The vaporization rate within the reboiler is the boilup V. Herein, the boilup V will generally be used as the manipulated variable associated with the reboiler.

Table 1.3 lists the manipulated variables in both the instrument context and the process context for the column illustrated in Figure 1.17. For the

TABLE 1.3. Manipulated Variables for Column in Figure 1.17

Instrument Context	Process Context
Distillate valve opening	Distillate flow D
Bottoms valve opening	Bottoms flow B
Reflux valve opening	Reflux flow L
Condenser cooling water valve opening	Heat transfer rate in condenser Q_C or overhead vapor flow V_C
Reboiler steam valve opening	Heat transfer rate in reboiler Q_R or boilup V

condenser and the reboiler, the manipulated variable in the instrument context depends on the equipment, but the manipulated variable in the process context does not.

1.10.5. Multivariable Control Problem

There are five controlled variables; there are five manipulated variables. This constitutes a 5×5 multivariable control configuration:

Controlled Variable	Manipulated Variable
Bottoms level	Distillate flow D
Reflux drum level	Bottoms flow B
Column pressure	Reflux flow L
Distillate composition	Overhead vapor flow V_C
Bottoms composition	Boilup V

This is only a listing of the controlled and manipulated variables; there is no significance to the order of either the controlled or manipulated variables.

In the single-loop approach to column control, a proportional–integral–derivative (PID) controller is configured for each of the controlled variables. The output of the controller must be to one of the control valves, but which one? The term "pairing" refers to selecting the manipulated variable to be used for each controlled variable. No pairing is implied in the previous list.

1.11. TOWER INTERNALS

The purpose of the tower internals is to facilitate mass transfer between the vapor and liquid phases within the tower. The options for tower internals are as follows:

Trays. Vapor–liquid contact is enhanced by dispersing the vapor into the liquid retained on the tray. The number of trays within each separation section is determined by the number of theoretical stages required by the design and the tray efficiency.

Packing. Liquid flowing over the packing provides a large wetted surface area for vapor–liquid contact. The height of each packed section is determined by the number of theoretical stages and the height of packing equivalent to a theoretical stage.

Trays versus packing is a tower design choice. Trays were primarily used in the older towers. Until the advent of structured packing, the maximum height of a packed section was restricted (the packing crushes under its own weight).

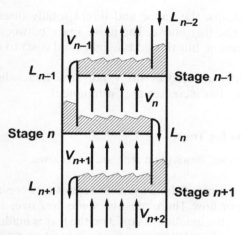

Figure 1.18. Vapor and liquid flows for trays.

But a packed tower tends to be smaller than a tray tower; in an existing tower, replacing trays with packing usually provides greater separation.

These and other design issues dictate the choice of trays versus packing. Control issues pertaining to trays versus packing are minor compared with the design issues.

1.11.1. Trays

Figure 1.18 illustrates the flows associated with trays. For a given tray, these are briefly as follows:

Liquid. Liquid flows from the tray above through a pipe called a down-comer. This liquid flows across the tray, then over a weir into the down-comer to the tray below.

Vapor. The vapor from the tray below enters through openings in the bottom of the tray and mixes with the liquid on the tray. The tray spacing provides for vapor–liquid disengagement so that, ideally, only vapor flows into the tray above.

Separation of liquid flow from vapor flow is not perfect:

Weeping. Some liquid "weeps" through the openings in the bottom of the tray, which to some extent short-circuits the vapor–liquid contact on the tray. Theoretically, valve caps and bubble caps prevent liquid weeping, but not in practice.

Entrainment. Ideally, the vapor and liquid totally disengage before the vapor enters the tray above. The more space between trays, the better the disengagement, but this adds height (and cost) to the tower.

Both are significantly affected by the vapor flow. Increasing the vapor flow reduces the weeping but increases the entrainment.

1.11.2. Vapor Flows for Trays

The limits on the vapor flows for a tray are as follows:

Minimum. The nature of the trays with regard to weeping establishes the minimum vapor flow. There must be some flow over the downcomer at all times. If not, the amount of liquid on the tray is inadequate and vapor–liquid contact is lost. If the vapor rate is too low, all liquid is lost from the trays, which rapidly increases the level in the reboiler. A shutdown on high reboiler level is likely to be initiated, which shuts off the heat to the column. But in any case, the result is a major upset to the tower.

Maximum. The vapor flow through the openings on the bottom of the tray results in a pressure differential across the tray. This pressure differential increases with the square of the vapor flow. If this pressure drop is too large, the consequence is a phenomenon known as "flooding." This is a major upset to the tower, so flooding will be examined in detail shortly.

The consequences of both high vapor flows and low vapor flows can be painful. Flooding normally receives the most attention. Most towers are equipped with pressure drop measurements that can detect the onset of flooding, and operations personnel take high pressure drops seriously. Unfortunately, there is no convenient measurement that can draw attention to low vapor flows, so its consequences are often a surprise to operations personnel.

1.11.3. Liquid Flows for Trays

The limits on the liquid flows for a tray are as follows:

Minimum. As noted above, some liquid must flow over the downcomer at all times. The loss of liquid through weeping to the tray below and entrainment to the tray above is largely offset by liquid gained through weeping from the tray above and entrainment from the tray below. Rarely would one attempt to operate a column at such low liquid flows that these factors would be significant.

Maximum. High liquid rates contribute somewhat to the pressure drop across a tray. This pressure drop occurs primarily at the point where the liquid flows from the bottom of the downcomer onto the tray. But unless

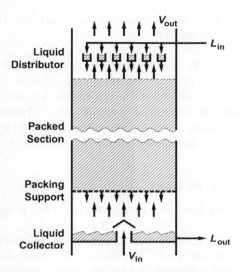

Figure 1.19. Vapor and liquid flows for packing.

this opening is unusually small, this contribution is minor as compared with the contribution from the vapor flow through the openings in the bottom of the tray.

High liquid rates are normally accompanied by high vapor rates; low liquid rates are normally accompanied by low vapor rates. For trays, the limiting conditions are normally attained due to the vapor flow, not due to the liquid flow.

1.11.4. Packing

Figure 1.19 illustrates the flows associated with a packed section. There are two items of equipment associated with each packed section:

Liquid distributor. Before entering a packed section, the liquid flows through a liquid distributor whose function is to distribute the liquid uniformly over the flow area of the packed section. For the upper packed section, the liquid flowing to the liquid distributor is the external reflux. For the lower packed section, the liquid flowing to the liquid distributor is the liquid from the upper packed section plus the liquid from the tower feed.

Liquid collector. The liquid flowing out of the packed section is collected by a liquid collector. For the upper packed section, the liquid from the liquid collector flows to the liquid distributor for the lower packed section. For the lower packed section, the liquid from the liquid collector is the liquid flow to the reboiler.

Packing comes in a variety of shapes and designs, most of which are proprietary. The objective is to provide the maximum amount of surface area for vapor–liquid contact per unit volume of the tower. A significant advancement was the introduction of structured packing, which permitted significantly greater heights of packed sections. The older packing was referred to as random packing and was basically dumped into the tower.

1.11.5. Vapor Flows for Packing

The limits on the vapor flows for a packed section are as follows:

Minimum. The vapor flow through a packed section can be stopped entirely. For example, start-up usually begins with the liquid flow. The vapor flow remains off until sufficient liquid has been admitted to the tower to completely wet the packing.

Maximum. The packing offers resistance to vapor flow, which results in a pressure differential across the packed section that increases with the square of the vapor flow. If this pressure drop is too large, the consequence is a phenomenon known as "flooding." The consequences of flooding on a packed section are the same as for trays.

As for tray towers, attention is directed to flooding. Most packed sections are equipped with pressure drop measurements that can detect the onset of flooding, and operations personnel take high pressure drops seriously.

1.11.6. Liquid Flows for Packing

The limits on the liquid flows for a packed section are as follows:

Minimum. If any vapor is flowing through a packed section, the liquid flow must be sufficient to keep the entire surface area of the packing wet with liquid. The consequences of a hot but dry packing surface are always adverse. The exact consequences depend on the nature of the materials being separated. In some cases, the consequence is a residue or buildup on the surface of the packing. For some materials, the dry surface is adversely affected (sometimes referred to as glazing) such that it will not be subsequently wetted by the liquid. The packing designers recommend what liquid flow is required to wet the packing.

Maximum. At high liquid rates, the flow area for the vapor is reduced, thus effectively increasing the pressure drop due to the vapor flow. However, high vapor rates usually accompany high liquid rates, and the adverse consequences of high vapor rates appear before any adverse consequences of high liquid rates.

For packed towers, logic is required within the controls to maintain an adequate liquid flow to keep the packing wet. However, no control logic is normally required to avoid high liquid flows for a packed section.

1.12. FLOODING

For all towers, the pressure is highest at the bottom of the tower and lowest at the top. This pressure drop is primarily a function of the vapor flow (actually vapor velocity); the contribution from the liquid flow is small.

Most towers are equipped with one or more differential pressure measurements, the options being the following:

Across the entire tower. Despite what some illustrations imply, the upper connection (the low pressure connection) is not always physically at the top of the tower. When the condenser and reflux drum are physically located at grade level, the low pressure connection is normally in the overhead vapor line. This connection can be at approximately the same physical elevation as the high pressure connection, which minimizes wet leg/dry leg issues that arise in differential pressure measurements.

Across a separation section. For these differential pressure measurements, the two connections will not be at the same elevation, so the wet leg/dry leg issues must be addressed. When the tower internals are the same in both separation sections, flooding initially occurs in the separation section with the highest vapor flow. Only the pressure drop across this separation section is required. But when the vapor flows are significantly different, the tower may have different tower internals or even different tower diameters (which affects the vapor velocity). In such cases, differential pressure measurements across both separation sections are recommended.

1.12.1. Pressures on Trays

Flooding is basically the same phenomenon for both tray towers and packed towers. Most find it easier to understand for trays, hence the illustration of a tray tower in Figure 1.20. The notation is as follows:

P_n = pressure on stage n (cm H_2O);
$\Delta P_n = P_n - P_{n-1}$ = pressure drop across stage n (cm H_2O);
H_n = height of liquid in the downcomer on stage n, relative to the weir (cm);
H_T = tray spacing (height between trays) (cm);
G = specific gravity of the liquid in the tower relative to water.

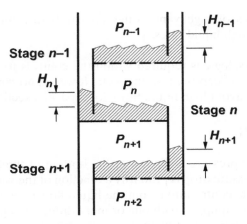

Figure 1.20. Pressures on trays.

The pressure on stage n is greater than the pressure on stage $n - 1$; that is, $\Delta P_n > 0$. In order for the liquid to flow from the downcomer to the tray, sufficient hydrostatic head is required in the downcomer to overcome this pressure differential. Consequently, the height of liquid in the downcomer depends on the pressure drop across the stage.

1.12.2. Hydrostatic Head in Downcomer

The height of liquid in the downcomer is given by the following expression:

$$H_n = G\,\Delta P_n.$$

The height of the liquid in the downcomer increases linearly with the pressure drop across the tray. But there is a limit on the available height in the downcomer. This limit is determined by the tray spacing; that is, the maximum allowable value for H_n is the tray spacing H_T. Consequently, there is a maximum allowable pressure drop for a stage:

$$\Delta P_n \leq G\,H_T.$$

If this pressure drop is exceeded, the liquid cannot flow from the downcomer onto the tray. This causes liquid to accumulate on the upper tray, which is said to "flood."

1.12.3. Contribution of Vapor Flow

The pressure drop across a tray is determined largely by the vapor flow, or more precisely, the velocity of the vapor as it flows through the orifices on the tray. This depends on both design and operational parameters:

Number and size of the openings in the tray. In towers where there are significant changes in the vapor flow between one separation section and another, the number and size of the openings may not be the same in both sections.

Vapor flow. The pressure drop increases with the square of the vapor flow. Logic is required in the control system to keep the vapor flow below its flooding limit.

Pressure. For a given vapor mass or molar flow, reducing the tower pressure increases the vapor velocity. Lowering the pressure in a tower could lead to flooding.

The tray spacing determines the maximum allowable pressure drop across a tray; this relationship is very simple and was presented previously. The maximum allowable pressure drop across a tray determines the maximum allowable vapor flow. However, accurately calculating the vapor flow that corresponds to the maximum allowable pressure drop is not generally possible.

1.12.4. Contribution of Liquid Flow

The hydrostatic head in the downcomer must also overcome any resistance to liquid flow. If the tray is designed properly, this resistance should be very small. The most likely location of any resistance is at the clearance at the bottom of the downcomer, which is illustrated in Figure 1.21.

There are two components to the hydrostatic head in the downcomer:

Vapor contribution. This is essentially $\Delta P_n/G$, with ΔP_n increasing with the square of the vapor velocity.

Liquid contribution. Assuming a proper clearance at the bottom of the downcomer, this component would only be significant at very high liquid flows.

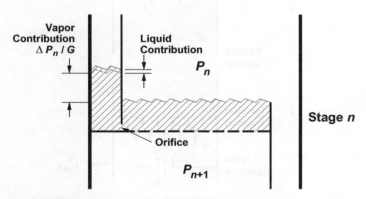

Figure 1.21. Contribution of liquid flow to tray pressure drop.

1.12.5. Maximum Pressure Drop for a Separation Section

Gravity provides the driving force for liquid to flow down a tower. The maximum driving force for liquid flow is the hydrostatic head provided by a column of liquid of the same height as the separation section. The maximum pressure drop for a separation section is

$$\Delta P_{max} = G\,H_S,$$

where

H_S = height of the separation section (cm);

ΔP_{max} = pressure drop across the separation section (cm H_2O);

G = liquid specific gravity relative to water.

It is essential that the pressure drop over a separation section never exceed this pressure drop.

The differential pressure measurement for a separation section or for the entire tower provides the basis for preventing flooding. Sometimes alarms are defined on the differential pressure measurement, and the operators are responsible for taking the appropriate action. However, it is also possible to incorporate logic so that the controls will take the necessary actions to avoid pressure drops that cause flooding.

1.12.6. Separation Sections with Packing

Figure 1.22 illustrates the pressure drop across a packed section. Flooding in a packed tower is basically the same as flooding in a tray tower. Gravity pro-

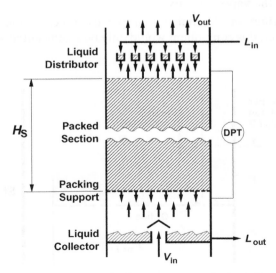

Figure 1.22. Pressure drop across a packed section.

vides the driving force for fluid flow down the tower. The packing provides resistance to vapor flow, which leads to a pressure drop across the separation section. As for tray towers, this pressure drop is proportional to the square of the vapor velocity. Should the pressure drop caused by the vapor flow exceed the hydrostatic head provided by a column of liquid of the same height as the packed section, liquid cannot flow down the packed section. Instead, it accumulates within the tower, which constitutes flooding.

Within the liquid phase in packed towers, there is very little resistance to fluid flow. This resistance would only be significant at very high liquid rates. Therefore, the major contribution to the pressure drop across a packed section is due to the vapor flow.

1.12.7. Issues Pertaining to Flooding

Flooding means the tower is filling with liquid. Consequently, reducing the liquid flow would seem to be an appropriate action to take. However, the tower is filling with liquid because the liquid cannot flow down the tower, not because too much liquid is being fed to the tower. The appropriate response to a flooding situation is to reduce the vapor flow. This is true for both tray and packed towers.

If (1) the tower is properly designed and (2) the tower is operating under the conditions for which it was designed, the limit on tower operations should be imposed by the most expensive component of the tower. In most cases, this is the tower internals. Consequently, encountering the flooding limit during production operations should be expected, and the controls must be configured accordingly.

The optimum operating point is often at a constraint. For a distillation column, this constraint is often associated with flooding. A subsequent chapter considers control configurations that will operate a tower close to the constraint imposed by flooding.

1.13. TRAY HYDRAULICS

Only those aspects of tray hydraulics that are of interest from a control perspective are examined herein. Specifically, the amount of liquid (the holdup) retained on a tray is affected by both the liquid flow and the vapor flow, the manner being as follows:

Liquid flow. The liquid holdup on a tray increases with liquid flow.

Vapor flow. The liquid holdup on a tray decreases with vapor flow.

This section examines the effect of the liquid flow; the next section examines the effect of the vapor flow.

1.13.1. Flow over Weirs

The flow of liquid over weirs has been extensively studied, one result being relationships between the height of liquid over a weir and the flow over the weir. These are routinely used in the water and wastewater industry, but their applicability to trays within a tower could certainly be questioned. Water flows in flumes and channels are relatively calm, whereas the liquid flowing across a tray in a tower is in a violent state of agitation. However, these relationships are used in various analyses, including tray efficiencies and tray dynamics.

The key relationship is the Francis weir formula, which is expressed as follows:

$$f = k \, w \, h^{3/2},$$
$$h = \left[\frac{f}{k \, w} \right]^{2/3},$$

where

h = height of liquid above weir (cm);
w = width of weir (cm);
f = volumetric flow (cc/s);
k = coefficient, 18.4 cm$^{\frac{1}{2}}$/s (3.33 ft$^{\frac{1}{2}}$/s).

1.13.2. Effect of Liquid Flow on Tray Holdup

On an increase in the liquid flow onto a tray, the height of the liquid above the weir must increase sufficiently so that the outlet flow is the same as the inlet flow. This means that some of the liquid flowing onto the tray is retained on the tray to cause the height over the weir to increase. On a decrease in liquid flow, the effect is the opposite. The effect is that of a first-order lag with the time constant being the hydraulic time constant τ_h.

The volume of liquid V_h above the weir is the product of the tray area A and the height above the weir h. Substituting the Francis weir formula for h gives a relationship for the effect of the liquid flow f on the volume of liquid above the weir:

$$V_h = A \, h = A \left[\frac{f}{k \, w} \right]^{2/3}.$$

The hydraulic time constant τ_h is the rate of change of the volume V of liquid above the weir with respect to the flow f:

$$\tau_h = \frac{dV_h}{df} = \frac{2A}{3 \, k \, w} \left[\frac{f}{k \, w} \right]^{-1/3}.$$

The hydraulic time constant depends on the following parameters:

Tray area A (a design parameter). The larger the tray area A, the larger the hydraulic time constant.

Length of the weir w (a design parameter). The longer the length of the weir w, the smaller the hydraulic time constant.

Flow f (an operating parameter). The hydraulic time constant is largest at low liquid flows.

1.13.3. Dynamic Effect of the Hydraulic Time Constant

The hydraulic time constant is manifested as a lag in the liquid flow within a separation section. For a single tray, the relationship between the outlet liquid flow and the inlet liquid flow is the characteristic first-order lag response. However, a separation section consists of some number of trays in series. The overall behavior of a large number of time constants in series is very similar to the behavior of dead time or transportation lag.

Figure 1.23 presents the response to a step increase in the liquid flow into a separation section that consists of 20 trays. The hydraulic lag on each tray is 6 seconds. The flow from tray 1 is the response of a 6-second lag to a step change in its input. The flow from tray 20 is the result of 20 lags in series, each lag being 6 seconds. This response is closer to the response of a dead time of 120 seconds (20 trays with a 6-second lag on each tray).

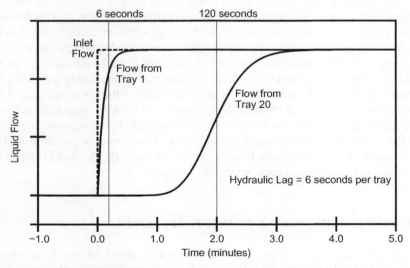

Figure 1.23. Effect of hydraulic lag on an increase in the liquid flow to a separation section.

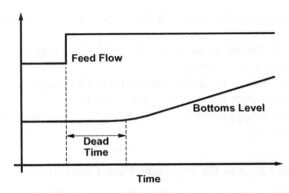

Figure 1.24. Response of bottoms level to increase in feed flow.

The hydraulic time constant is typically in the range of 5–10 seconds, but since internal flows cannot be measured, an accurate value is not generally available.

1.13.4. Response in Reboiler Level

The consequences of the hydraulic time constant are often observed in the response of the bottoms level to changes in either the feed flow or reflux flow. The ensuing discussion assumes all controls are on manual, so the response will be that of the tower alone.

If the feed is mostly liquid, an increase in the feed rate means an increase in the liquid flow to the lower separation section of the tower. This leads to an increase in the liquid flow out of the lower separation section and into the reboiler. With no controls in operation, this causes the reboiler level to increase.

Figure 1.24 illustrates the response of reboiler level to an increase in the feed rate. The reboiler level does increase, but not immediately following the increase in the feed rate. There is a delay, after which the reboiler level increases in the expected manner. The value for the delay, often referred to as dead time, is determined by the number of trays in the lower separation section and the hydraulic time constant of each tray. If the lower separation section has 10 trays and the hydraulic time constant is 6 seconds, the dead time in the reboiler level response is approximately 1 minute.

1.14. INVERSE RESPONSE IN BOTTOMS LEVEL

On an increase in the heat input to a reboiler, the bottoms level should decrease. Indeed, this is always the long-term effect; however, the short-term effect can be that illustrated in Figure 1.25. The bottoms level initially increases,

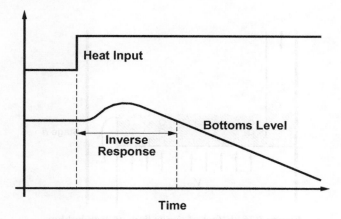

Figure 1.25. Inverse response in bottoms level.

but eventually decreases as expected. This type of behavior is known as "inverse response."

Towers exhibit inverse response to varying degrees:

- Some exhibit little or none. Given the noise generally present in level measurements associated with a boiling liquid, a small amount of inverse response would be difficult to detect. Any smoothing on the level measurement would also obscure the inverse response.
- Some exhibit what appears to be dead time; following the increase in the heat input, some time elapses before the bottoms level begins to drop.
- Some exhibit inverse response to a minor degree.
- Some exhibit inverse response to a very noticeable degree. Such a tower was reported by Buckley et al. [2].

Inverse responses can be extreme. However, few, if any, columns exhibit inverse response to this degree.

1.14.1. Effect of Vapor Flow on Tray Holdup

Vapor enters the tray through small openings and is dispersed into the liquid on the tray. If the liquid on the tray is in a quiescent state as illustrated in Figure 1.26, the volume of liquid on the tray is the tray volume (tray area times weir height) less the volume of liquid displaced by the vapor bubbles.

What happens when the vapor flow increases? More vapor bubbles are dispersed into the liquid, displacing a greater volume of liquid. The trays are said to "dump liquid." In the tower, an increase in vapor flow is felt on all trays in the tower. Therefore, each tray is "dumping liquid" into the downcomer and

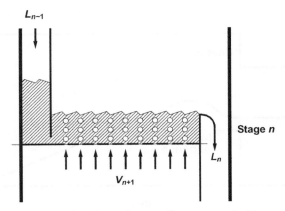

Figure 1.26. Effect of vapor flow on tray holdup.

onto the tray below. All of this liquid eventually ends up in the tower bottoms, resulting in an increase in bottoms level.

This can lead to an interesting sequence of events. If one increases the heat input to the reboiler, the expectation would be for the vapor flow to increase and the bottoms level to drop. This is indeed the long-term effect. However, the short-term effect is potentially quite different. The increase in vapor flow causes the trays to dump liquid, which causes the bottoms level to increase. The initial response is in a direction opposite of the long-term response, hence resulting in what is known as an "inverse response."

In columns that exhibit dead time in the response of bottoms level to an increase in heat input, there is no true transportation lag. But during what appears to be the dead time, the decrease in bottoms level due to the increased boilup is basically offset by the liquid being dumped by the trays.

1.14.2. Mean "Liquid" Density

The liquid on a tray is in a rather violent state of agitation caused by the dispersion of vapor bubbles into the liquid. The concept of a quiescent pool of liquid on a tray is not accurate (nor is the concept of liquid calmly flowing over a weir).

Consider the vapor–liquid mixture on the tray to be the "liquid phase." This mixture generally extends well above the weir height, but does not extend to the tray above (should it extend to the next tray up, significant liquid entrainment would occur and the tray efficiency drops dramatically). What is the effect of vapor flow on the mean density of the "liquid phase" (in reality, the vapor–liquid mixture)?

Most information on the effect of the vapor flow on the mean "liquid phase" density comes from tray efficiency studies. At low vapor flows, changes in the

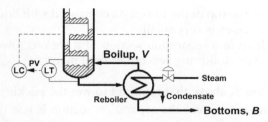

Figure 1.27. Control bottoms level with boilup.

vapor flow affect the mean density and consequently the liquid holdup on the tray. But at high vapor flows, changes in the vapor flow have little effect on the mean density and the liquid holdup. This suggests that towers with low vapor flow rates would exhibit a more pronounced inverse response than towers with high vapor flow rates.

1.14.3. Bottoms Level

If the bottoms level exhibits inverse response, will the level control loop illustrated in Figure 1.27 deliver satisfactory performance? When the level controller increases its output, the expected response is a decrease in bottoms level. But when inverse response is present, the short-term result is an increase in the bottoms level, which causes the controller to further increase its output.

Inverse response always has a negative impact on the performance of a loop, even more than dead time. In the presence of dead times, controller gains must be reduced. Inverse response necessitates even lower values of the controller gain. The inverse response in bottoms level is generally mild to at most moderate. While level control performance suffers, the result is usually acceptable.

1.15. COMPOSITION DYNAMICS

For the same number of stages, the separation provided by a tray tower and by a packed tower is exactly the same. Whether the tower is trays or packed has little impact on the steady-state solution. However, this is not the case for the dynamics.

1.15.1. Vapor and Liquid Dynamics

The vapor dynamics are the same for both tray and packed towers. Vapor dynamics are also the simplest: any change in vapor flow is propagated instantly throughout the tower. That is, a change in the vapor flow from the reboiler is immediately felt at the top of the tower. As compared with liquid holdups, the vapor holdup in a tower is very small.

The liquid holdup in a separation section of a packed tower is essentially constant provided the following two criteria are met:

- The liquid flow is above that required to wet the packing.
- The pressure drop across the separation section is less than about 80% of the flooding limit.

A constant liquid holdup means that a change in the liquid flow into the separation section immediately appears in the liquid flow out.

The liquid holdup on trays is far more complex. The liquid holdup increases with the liquid flow, but decreases with the vapor flow. Both of these were examined previously, so no further discussion is required.

1.15.2. Composition Dynamics

The composition dynamics are largely determined by the liquid holdups within the tower. Liquid holdups include the following:

Tower internals. The liquid holdup in a tray tower is generally larger than the liquid holdup in a packed tower.

Reflux drum. A few towers do not have a reflux drum. In small vacuum towers, the condenser is often physically located at the top of the tower, with the condensate returned directly to the upper separation section.

Bottoms holdup. This depends on the type of reboiler. For thermosyphons, it is in the bottom of the tower; for kettle reboilers, it is in the reboiler itself. A later section examines various types of reboilers.

A dynamic simulation must encompass all of these. Since the liquid holdup is less in a packed tower, the contribution of the condenser and reboiler to the overall tower dynamics is larger for a packed tower than for a tray tower.

1.15.3. Stage Dynamics

The simplest approach to simulating stage dynamics is to assume that the liquid within the stage is perfectly mixed. This assumption is generally made for both packed and tray towers.

For packed towers, this assumption is clearly not correct. Theoretically, there should be no composition gradients in the horizontal direction. However, there is little vertical mixing in the liquid within a packed tower, so vertical gradients are present within the stage. Such composition gradients can be simulated, but do the improved results justify the extra effort? Assuming that the liquid in a stage is perfectly mixed permits the simulation program for a tray tower to be also used for a packed tower.

In developing the equations for the steady-state simulation, the liquid on a tray is assumed to be perfectly mixed. Would the liquid on a tray in a 10-m-diameter tower be perfectly mixed? Of course not. For the steady-state simulations, the tray efficiency compensates for the errors. For the dynamic simulation, assuming the tray consists of two perfectly mixed sections would give slightly different results than assuming the entire tray is perfectly mixed. However, the results are not drastically different. Dynamic simulations can be developed for any assumption regarding the mixing on the tray, but what assumption should be made? Until this question can be answered, assuming the entire tray is perfectly mixed will continue.

1.15.4. Dynamic Simulations

Before undertaking a dynamic simulation, a steady-state simulation is a must. But a dynamic simulation requires many additional parameters. For example, stage holdups and reflux drum capacity have no effect on a steady-state simulation, but must be known for a dynamic simulation. The additional parameters include the following:

Flows. Steady-state simulations can be done on the basis of a feed rate of 100 mol/h. However, dynamic simulations require actual flow rates.

Tower size. The dynamic simulation requires the number of actual stages, the column diameter, column height, and so on.

Tower internals. To determine the liquid holdup for each stage, the nature of the separation sections must be known.

Condenser. At a minimum, the capacity of the reflux drum must be known. The type of condenser must be known, and for some, the size must also be known.

Reboiler. The type of reboiler must be known. The capacity of the liquid holdup in the bottoms must be known, and for some types of reboilers (e.g., kettle reboilers), their size must also be known.

Programs to simulate column dynamics are widely available. However, much effort is required to obtain the additional parameters.

1.15.5. Simulation Detail

The degree of detail for a simulation must be consistent with the intended use of the simulation. The greater the degree of detail, the greater the number of parameters that will be required.

The simplest situation is when the primary requirement is to simulate the composition or temperature dynamics. The purpose for such a simulation may be to verify that the temperature or composition control configuration will function properly. The composition and temperature loops are the slowest loops. Such simulations can be simplified by assuming that the level loops, the flow loops, and the column pressure loop are much faster. Assuming perfect performance from a loop has several attractive consequences:

- A constant value to be used for its controlled variable (the level, flow, or pressure).
- Values of several parameters (measurement device characteristics, control valve characteristics, etc.) are not required.
- The flow through the control valve can be calculated from a steady-state equation, thereby eliminating additional parameters.

Compared with the temperature and composition loops, flow loops will be instantaneous. The column pressure is very likely to be instantaneous (except possibly for a condenser arrangement known as the "flooded" condenser).

Assuming constant level deserves somewhat more attention. Using the reflux drum as the example, there are two possibilities:

Control level by adjusting the distillate flow. With perfect level control, any change in the overhead vapor flow is immediately translated to a change in the distillate flow. As the distillate flow is to external equipment, errors in this assumption would not significantly affect the results of the dynamic simulation.

Control level by adjusting the reflux flow. With perfect level control, any change in the overhead vapor flow is immediately translated to a change in the reflux flow. As the reflux flow is returned to the tower, significant errors in this assumption would affect the results of the dynamic simulation.

Similar issues arise for the reboiler.

1.15.6. Dynamic Simulation as Part of Design

Most process designs are based largely on steady-state relationships. Arguments have been advanced for dynamic simulations to become an integral part of process design. The progress, if any, has been slow.

Driven by financial reasons, shortening the design and construction cycle is of major interest. One consequence is that major items of equipment must be ordered as early as possible, often before all aspects of the design are finalized. Distillation columns are usually major items of equipment.

One claim for dynamic simulation is that it will uncover flaws in the design. Experience seems to support this claim. But to take advantage of this for a major item of equipment, placing the order must be delayed to provide the time to make the simulation. But this is counter to the objective of shortening the design cycle. Doing the simulation in parallel is an option. However, the simulation must be delayed until detailed designs are available for the condenser, the reboiler, and the tower internals. Undertaking the simulation at that time is too late; by the time the results are available, the equipment fabrication and plant construction are too far along.

REFERENCES

1   Yaws, C. L., *Chemical Properties Handbook*, McGraw-Hill, 1999.
2   Buckley, P. S., R. K. Cox, and D. L. Rollins, Inverse response in a distillation column, *Chemical Engineering Progress*, 71(6), June 1975, 83–84.

2

COMPOSITION CONTROL

Product specifications state composition or some property that is a function of the composition. This suggests that the composition of the product should be directly measured, which entails a composition analyzer. The pros and cons are quite simple:

- The analyzer reports the composition of the product stream (or a stream nearby).
- Analyzers (along with their sample systems) are expensive to purchase, install, and maintain.

An alternative is to use temperature measurements. These are less expensive and more reliable than analyzers, but many issues arise.

This chapter examines various issues pertaining to the composition of a product:

- Product specifications
- Analyzers versus temperature
- Single-end control configurations for
  - Distillate composition only
  - Bottoms composition only
- Impact of reflux drum level control on distillate composition control

Double-end composition control is the subject of a subsequent chapter.

Distillation Control: An Engineering Perspective, First Edition. Cecil L. Smith.
© 2012 John Wiley & Sons, Inc. Published 2012 by John Wiley & Sons, Inc.

2.1. PRODUCT SPECIFICATIONS

A distillation column separates the feed into two (or more) product streams, one or more of which must meet specifications. The operational targets derived from the product specifications fall into two categories:

Composition. The composition target may be on a single component or on some combination of components (e.g., total impurities).

A property that is a function of composition. This includes heating value, density, octane number, penetration number for asphalt, and odor.

2.1.1. Properties and Compositions

A given composition for a product will give a certain value for a property. But for multicomponent systems, there will be many compositions that will give a certain value for that property. The situations include the following:

- Some properties can be precisely computed from the composition. For example, the heating value of natural gas can be computed from the gas composition and the heating values of the components. At one time, calorimeters were mandated for measuring natural gas heating values, but these have largely disappeared.

- Some properties can be approximately computed from the composition. Control actions can be based on the computed values of the properties, but periodic tests are required either to verify that the computation gives the correct result or to adjust the computed value of the property.

- Some properties cannot be computed from the composition. The penetration number for asphalt is determined by dropping a metal ball and measuring the depth of its penetration into the asphalt. Computing such a property from composition is challenging, and furthermore, obtaining a composition analysis for asphalt is also challenging.

- The property is potentially affected by trace amounts of impurities. Odor is such a property. A composition analysis can be performed for the major constituents and for certain impurities. But the final product must pass a sniff test by a human nose.

2.1.2. One-Sided Targets

When the product specification is a composition, a single target value is usually specified. The specification may pertain to either of the following:

One or more impurities. The target becomes the maximum allowable level of the impurities. However, any lesser value for the impurities is acceptable.

Product purity. The target becomes the minimum acceptable product purity. However, purities in excess of this value are acceptable.

Of these two, specifications pertaining to the impurities in the product are more common.

The above statements pertain only to the acceptability of the product. In most cases, removing more impurities than required to meet the product specification entails costs, usually energy but sometimes column throughput or recovery. Therefore, the most common operational objective for a column is to operate such that the product meets the specifications but does not exceed the specifications by an unreasonable amount.

2.1.3. Two-Sided Targets

These are more commonly encountered when the specification pertains to a physical property. A range is specified for the acceptable values for the physical property. The range is defined by two values, one being the minimum acceptable value for the property and the other being the maximum acceptable value for the property.

Such specifications are intended to define the characteristics that a product must possess to be suitable for a specific application. Viscosity is a physical characteristic that is crucial to some applications. A range is almost always specified for viscosity. If the viscosity is too low, the product flows too freely and is not acceptable. If the viscosity is too high, the product flows too slowly and is not acceptable.

For most columns, cost issues make operating at one end of the acceptable range more desirable than operating at the other end. Rarely is operating somewhere within the range more economically attractive than operating at one of the ends.

2.1.4. Multiple Characteristics

Table 2.1 presents the **Gas Processors Association (GPA)** standard 2140 for propane (HD-5). Not only are individual characteristics stated, but the test methods are also stated. This raises issues pertaining to on-line analyzers— are the analytical results acceptable in lieu of the stated test method? Not always. Operations rely on the results from the on-line analyzer to make adjustments to process conditions, but the final acceptability of the product must be determined by off-line test procedures in accordance with the specified test method.

While specifications such as in Table 2.1 add some complexity to the production operations, they also provide opportunities. All impurities in the propane product are sold at the price of propane. This makes the following possibilities of interest:

TABLE 2.1. Product Specifications for Propane (HD-5)

Product Characteristic	Target	Test Method
Composition (liquid vol%)		
Ethane, max	2.0	ASTM D-2163
Propane, min	96.0	ASTM D-2163
Butane, max	2.5	ASTM D-2163
Pentanes+, max	0.1	ASTM D-2163
Unsaturates, max	0.1	ASTM D-2163
Vapor pressure at 37.8°C, kPag, max	1379	ASTM D-2598
Corrosive compounds copper strip	Pass	ASTM D-1838
Total sulfur, ppm, max	185	ASTM D-2784

- If the product value of ethane is less than the product value of propane, then put as much ethane in the propane as the specifications allow.
- If the product value of butane is less than the product value of propane, then put as much butane in the propane as the specifications allow.

The depropanizer will not be able to influence all of the characteristics stated in Table 2.1. As noted previously, essentially all of the ethane in the feed to the depropanizer will leave with the propane product. One must examine the entire production facility to determine where each of the individual characteristics in the product specification can be controlled.

2.1.5. Units for Compositions

The stage-by-stage separation models always compute the molar composition (mol% or mole fraction) for each product stream. However, rarely are product specifications explicitly in mol%:

Liquids. The composition is usually either wt% or vol%. For a multicomponent system, a specification such as wt% total impurities cannot be converted to a mol% total impurities. Instead, the wt% total impurities must be computed from the molar composition of the stream.

Gases. The composition is normally stated as vol%. For ideal gases, this is mol%.

Knowing the liquid specific gravity and the molecular weights, the results of a composition analysis of the propane product stream can be converted to whatever units are required. For example, suppose the product is 2% ethane, 96% propane, and 2% butane by liquid volume. The composition in the other units is as follows:

Component	Specific Gravity	Molecular Weight	Volume (%)	Weight (%)	Mole (%)
Ethane	0.315	30	2.0	1.3	1.9
Propane	0.493	44	96.0	96.4	96.6
Butane	0.573	68	2.0	2.3	1.5

A total stream analysis can always be converted between vol%, wt%, and mol%. However, product specifications rarely state the total stream analysis. The specification for propane is typical—a maximum or minimum is specified for selected components in the product.

For example, the specifications for propane state that the composition of propane must be at least 96% by liquid volume. How can this be converted to wt% or mol%? To do so exactly, the composition of the impurities must be known. The following illustrate the effect:

Maximum ethane, maximum butane. The 4 vol% impurities would have to be split almost evenly between ethane and butane (2 vol% ethane, 96 vol% propane, 2 vol% butane). The product would be 96.4 wt% propane or 96.6 mol% propane.

Maximum ethane, minimum butane. The composition should be 2 vol% ethane, 98 vol% propane, and no butane. The product would be 98.7 wt% propane or 98.1 mol% propane.

Minimum ethane, maximum butane. The composition should be 2.5 vol% butane, 97.5 vol% propane, and no ethane. The product would be 97.1 wt% propane or 98.1 mol% propane.

While the changes in the values for wt% or mol% are small, they are not insignificant in towers producing commodity products (small changes are spread over a large production volume to give a significant result).

A further complication is that the same component can appear in more than one specification. Consider ethane:

- A value is explicitly stated for the maximum amount of ethane.
- The specification of 96% propane minimum is equivalent to 4% total impurities maximum. Ethane is an impurity.
- Ethane contributes to the vapor pressure for the propane product.

In the final product, each specification can be applied. But as will be discussed shortly, the depropanizer cannot influence the amount of ethane in the propane product—whatever ethane is in the feed to the depropanizer will flow to the propane product. The feed to the depropanizer is often the bottoms stream from a deethanizer. For controlling the deethanizer, the target for the bottoms product is often the molar ratio of ethane to propane. All specifications relating to the ethane in the propane product must be taken into consideration to establish such a target.

2.2. COLUMNS IN SERIES

Figure 1.13 illustrates a separation train consisting of four columns in series—a demethanizer, a deethanizer, a depropanizer, and a debutanizer. Herein the focus will be on the compositions that must be controlled for the depropanizer. However, similar issues arise for the other towers.

Only the following three aspects of the specifications in Table 2.1 for the propane product will be considered:

- The product must be 96% or more propane. Herein an alternate statement will be used: The total impurities in the propane product must not exceed 4%.
- The product must be 2% or less ethane.
- The product must be 2.5% or less butane.

The other specifications will be assumed to be either met or exceeded. The safe approach to operating the separation train is to drive all impurities in all product streams to their minimum possible values. One downside is energy costs. To reduce the level of any impurity requires enhanced separation, which in an existing tower means more energy. Another downside pertains to the value of each impurity relative to the value of the product stream in which they appear.

2.2.1. Optimum Amounts of Impurities in Product Streams

In the separation train in Figure 1.13, the following cases relate in some way to the depropanizer:

Amount of butane in the propane product stream. This is determined by the depropanizer.

Amount of propane in the butane product stream. This is also determined by the depropanizer. The bottoms stream from the depropanizer is the feed to the debutanizer.

Amount of ethane in the propane product stream. This is determined by the deethanizer. The bottoms stream from the deethanizer is the feed to the depropanizer.

The optimum level of an impurity in a product stream always depends on two product values:

Value of the product stream that contains the impurity. For the depropanizer, the distillate stream is the propane product.

Value of the impurity. This is the value of the product stream that consists primarily of the impurity. The major impurities in the propane product stream are ethane and butane. The value of the ethane impurity in the propane is the value of the ethane product stream. The value of the butane impurity in the propane is the value of the butane product stream.

Optimizing the impurity levels involves the following trade-offs:

- Removing 1 kg of ethane from the propane product stream increases the ethane product stream from the deethanizer by approximately 1 kg.
- Removing 1 kg of butane from the propane product stream increases the butane product stream from the debutanizer by approximately 1 kg.

The respective values result in two cases:

The value of the product stream that contains the impurity is greater than the value of the impurity. The optimum amount of the impurity is the maximum amount imposed by the product specifications. For example, the product specifications for propane permit up to 2.5 vol% butane. This butane is sold at the price of propane. Why spend money (in the form of energy) to remove butane from the propane product, only to get less for the butane when it is sold as butane?

The value of the product stream that contains the impurity is less than the value of the impurity. It may be economically beneficial to lower the level of the impurity below what the specifications permit. The optimum balances the following:

- The increased energy costs required to lower the impurity level.
- The increased return from selling the impurity at the value of its corresponding product stream.

For example, lowering the level of butane in the propane product stream requires additional energy. But additional revenue is generated because butane in the butane product stream is sold at the price of butane, whereas the butane in the propane product stream is sold at the price of propane.

An incremental formulation of the latter optimization problem will be presented in a subsequent chapter.

2.2.2. Composition Controls

When designing the composition controls for a distillation column, the composition to be controlled must be one that is affected to a significant degree by the controls. The controls primarily influence two compositions:

The heavy key in the distillate product. For the depropanizer, this would be the butane in the propane product.

The light key in the bottoms product. For the depropanizer, this would be the propane in the feed to the debutanizer.

The controls for a column have little (or no) influence on the off-key components:

- The lighter-than-light-key components in the feed appear almost entirely in the distillate product. For the depropanizer, this includes the ethane in the propane product. This composition appears in the propane specifications; however, it cannot be controlled at the depropanizer.
- The heavier-than-heavy-key components in the feed appear almost entirely in the bottoms product. For the depropanizer, this includes the pentane in the feed to the debutanizer.

From the perspective of control, the Hengstebeck approximation is applicable.

For the separation train in Figure 1.13, the impurities associated with the depropanizer must be controlled as follows:

Ethane in propane. This is determined by the composition of the bottoms stream from the deethanizer.

Butane in propane. This is determined by the depropanizer.

Propane in butane. This is determined by the composition of the bottoms stream from the depropanizer.

For the depropanizer in the separation train in Figure 1.13, controls are required for both key components:

Butane in the distillate product. This must be controlled so as to meet the specifications for the propane product.

Propane in the bottoms product. This must be controlled so as to meet the specifications for the butane product.

2.2.3. Targets

Ultimately the target values for each stream are determined from the product specifications. Sometimes the specifications can be used directly as a target, but sometimes a derived target must be used. This is the case for the depropanizer:

Distillate product. The product specifications pertaining to butane in the propane product can be directly used by composition controls on the distillate stream from the depropanizer.

Bottoms product. The product specifications pertaining to the propane permitted in the butane product cannot be used directly as the target for propane in the bottoms product from the depropanizer. This stream contains pentane and higher molecular weight components that will not be present in the butane product.

Suppose the butane product is to be 98% butane, 2% propane, and negligible amounts of other impurities. The ratio of propane-to-butane in the butane product is 1:49. This same ratio would be required in the bottoms of the depropanizer provided the following assumptions are valid:

All of the propane in the feed to the debutanizer leaves with the butane product. Propane is a lighter-than-light key in the debutanizer, so this assumption is justified.

All of the butane in the feed to the debutanizer leaves with the butane product. Butane is the light key in the debutanizer, so some butane will be in the bottoms product from the debutanizer. This is a small amount, but is not zero.

If desired, the propane-to-butane ratio in the bottoms of the depropanizer can be adjusted to compensate for the butane recovery in the debutanizer. However, such recoveries are normally high, so the adjustment would be small.

2.3. COMPOSITION ANALYZERS

Analyzers that are potentially capable of providing a composition analysis of a multicomponent product stream from a distillation tower include the following:

- Gas chromatograph
- Liquid chromatograph
- Infrared (IR) spectrometer
- Near-infrared (NIR) spectrometer
- UV spectrometer
- Mass spectrometer
- Chemoluminescence analyzer
- Nuclear magnetic resonance (NMR) analyzer

Occasionally, one encounters binary systems, where technologies such as density, refractive index, and electrical conductivity, can be applied. However, these tend to be the exception.

Of the above technologies, the gas chromatograph is most commonly applied to distillation columns.

2.3.1. Continuous versus Sampling

Analyzers can be classified as follows:

Continuous. The analyzer produces a continuous output for the component of interest.

Sampling. A sample is injected into the analyzer and an analysis is produced for that sample. The time interval on which the sample is injected is known as the sampling time or sampling interval.

The gas chromatograph is a sampling analyzer. A sample is injected into the analyzer, and some time later (the analysis time), the results are reported. Usually, the sampling time is only slightly longer than the analysis time.

Occasionally, a very expensive analyzer is multiplexed to serve several columns. If an analyzer with a 15-minute sampling time is multiplexed to serve four columns, the sampling time for a given column will be 1 hour.

Sampling always has a negative impact on the performance of the controls. The effect is nil if a fast analyzer (one with a short sampling time) is installed on a slow column. For a natural gas sample, a chromatograph can produce a total composition analysis in about 15 seconds. However, a total composition analysis of a heavy oil may take 15 minutes. As will be discussed shortly, this needs to be quantified along with some other factors.

2.3.2. Sample Points

On the process schematic diagrams presented herein, the analyzers are indicated as being on the product streams. Product specifications pertain to the product streams, so this is indeed the composition of interest. However, the sample points are often not on the product stream itself.

The location of the sample point is a compromise between how close it is to the product stream and the quality of the sample. The sample for a gas chromatograph is preferably withdrawn from a vapor stream instead of a liquid stream (completely vaporizing liquid samples is problematic). The consequences are as follows:

Overhead vapor. For a partial condenser, the distillate product is a vapor stream from which the sample can be withdrawn. For a total condenser, the sample can be withdrawn from the overhead vapor stream. The steady-state composition of the distillate will be the same as the composition of the overhead vapor stream.

Reboiler vapor. At the bottom of a tower, the sample is preferably free of the high molecular weight components ("gunk") that are often present. The sample can be withdrawn from the boilup provided by the reboiler, or for tray towers, the sample can be withdrawn a couple of trays up the column. The relationship between the composition of the sample and the composition of the bottoms stream is determined by the vapor–liquid equilibrium relationships.

2.3.3. Analyzer Location

The physical location of analyzers within a production facility is not a trivial issue. One option is to merely protect the analyzers from wind and rain, leaving them exposed to variations in ambient temperature, the atmospheres in industrial facilities (which may be hazardous), and so on. The other option is to physically locate the analyzers in an "analyzer house" that is specifically constructed to house the analyzers. This has the following advantages:

Conditioned space. The specifications on temperature limits for process analyzers have expanded, some claiming to operate from 10°C (50°F) to 50°C (122°F). Noncondensing is also usually a requirement. However, analyzers are complex items of equipment that seem to work best around 20°C (68°F). Analyzer technicians always prefer to work in conditioned space.

Nonhazardous atmosphere. Intrinsically safe models are now available for some analyzers. However, long exposures to low levels of hydrogen sulfide and the like will impact any complex item of electronic equipment.

Utilities. Analyzers require power, instrument air, various gases, water, and so on. Either the analyzers are physically located where all such services are available, or the required services must be routed to the location of each analyzer.

Sample systems. Sample systems require routine monitoring on the part of the operations staff. This is easier when the sample systems terminate at a single location.

2.3.4. Sample System

From an operations perspective, perhaps the major problem with analyzers is that most require a sample system. The success or failure of an analyzer installation often depends on the performance of the sample system. Supply the analyzer with a clean sample and it will do what it is designed to do. However, one "burp" from the sample system will put most analyzers out of commission.

The design of sample systems for analyzers proves to be extremely challenging. A representative sample must be extracted from the process, transported to the physical location of the analyzer, conditioned as required by the analyzer, and finally returned to the process (or suitable alternate destination). The design must utilize small equipment to deliver consistent performance

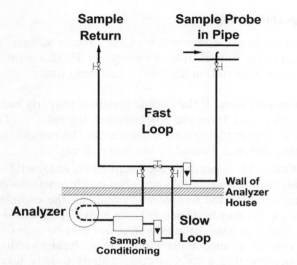

Figure 2.1. Analyzer sample loop.

with acceptable maintenance requirements. The general design illustrated in Figure 2.1 contains two "loops":

Fast loop. The purpose of this loop is to transport the material to be analyzed from the process to the physical location of the analyzer in a reasonable time. Where the distances are large (such as in refineries), the pipe sizes are on the order of 2 or 3 cm. Because of the large volumes flowing around this loop, the piping contains only valves and a flow indicator—no sample conditioning is provided. For gas samples, there must be no condensation anywhere in this loop, so heat tracing is usually required. If an analyzer house is provided, these pipes are entirely outside of the analyzer house.

Slow loop. The tubing is of small size and the flow is low. To avoid transportation lag, the physical distances in the slow loop must be short. The sample conditioning required to provide a suitable sample to the analyzer is incorporated into the slow loop. The dynamic characteristics of this equipment are crucial. For example, centrifugal separators (cyclones) are preferred over knockout vessels because of a faster dynamic response; the downside is that centrifugal separators require a larger pressure drop.

Unfortunately, proceeding from generalities to specifics can be challenging. The design of a sample system is a special technology. Prior experience with the materials being analyzed is very desirable, else there is likely to be a learning curve.

2.3.5. Transportation Lag

A change occurs in the composition of the process stream. When will this change appear in the output from the analyzer? There are three contributors to the delay or transportation lag (also called dead time):

Sample transport time. If the sample system is properly instrumented, this can be calculated. From the flow indicator, the velocity of the fluid in the sample transport piping can be calculated. The sample transport time is the piping distance divided by the flow velocity.

Analysis time. This depends on the nature of the analyzer. For IR analyzers, the analysis time is essentially zero. But for chromatographs, it is the time between injecting the sample and obtaining the complete results. For light gases, this may be 15 seconds; for heavy oils, it could be 15 minutes. If this time is excessive, the manufacturer may be able to make changes in the column arrangements within the analyzer to reduce the time. In addition, obtaining a total stream analysis usually takes longer than obtaining the ratio of two components.

Effect of sampling. An analyzer that samples on an interval of 1 minute contributes about 0.5 minute to the transportation lag. Sometimes the sample is taken just after the process change occurs, and no time is lost. Sometimes the sample is taken just before the process change occurs, and one sampling time is lost. On the average, the loss is half the sampling time.

The impact of the total dead time on loop performance depends on the dynamics of the tower itself. A crude characterization of the tower dynamics uses a process dead time and a process time constant. If the total dead time contributed by the analyzer and its sample system is less than half of the process dead time, little benefit is realized by further shortening the contribution from the analyzer and its sample system. Unfortunately, values for the process time constant and process dead time are rarely readily available.

2.3.6. The Case for Analyzers

The case for analyzers is really very simple—they tell you what you need to know:

Product specifications are in terms of compositions (or variables related to composition). Consider the propane specifications. The minimum value for the vol% propane, the maximum value for vol% ethane, and the maximum value for vol% butane were specified explicitly. Another specification was the maximum vapor pressure at 37.8°C. Knowing the stream composition, the vapor pressure at 37.8°C can be computed from the vapor–liquid equilibrium relationships.

Analyzer is on or near the product stream. The sample may not be drawn from the product stream, but rarely is it more than a stage or two removed.

Less frequent quality control analyses in the lab. This is controversial. Notice "less frequent"; the complete elimination of analyses in the quality control (QC) lab is not assured. Furthermore, resolving differences between the process analyzer and the QC lab results can be challenging.

2.3.7. The Case Against Analyzers

The case against analyzers largely involves costs:

Analyzers are expensive to install. The cost of analyzers has always been high, but as we proceed to ever more sophisticated analyzers, the costs continue to escalate. And the installation cost must also include the sample system.

Analyzers are high maintenance items. The level of effort is often stated to be 4 hours of a technician's time per analyzer per week. In addition, checking the flows, pressures, and other conditions within the analyzer sample system must be a routine part of plant operations. To date, the general practice is to equip analyzer sample systems with local indicators, not transmitters.

Analyzers require specialized technicians. Distillation applications require some appreciation of the principles of vapor–liquid equilibrium. For example, if the temperature of a cylinder containing a standard sample for calibration is allowed to drop below a certain value, internal condensation will occur and the gas composition is affected. The technicians must understand this.

2.4. TEMPERATURE

No product specifications state a stage temperature. Product specifications always state compositions or properties that are functions of composition.

In a certain sense, a stage temperature is being used to infer a stream composition. However, the approach is really more elementary than this. By maintaining a constant value for a stage temperature, the composition of the product stream will hopefully remain essentially constant. Unfortunately, this is not assured. Especially for multicomponent systems, the relationship between stage temperature and product stream composition is surprisingly complex:

- The stage temperature is strongly affected by pressure.
- Variations in off-key components affect the stage temperature.

- Low sensitivities and nonlinear characteristics present problems for the controls.
- Proper location of the stage for the temperature measurement is crucial but is not assured.

2.4.1. Temperature Control Stages

When a stage-by-stage separation model is used as the basis for the column design, the computed values include the temperature on each stage. Plotting these temperatures gives the temperature profile presented in Figure 1.15.

There can be a temperature control stage in each separation section:

Upper control stage. This stage is in the upper separation section. By maintaining a constant value for the temperature of this stage, the distillate composition hopefully remains essentially constant.

Lower control stage. This stage is in the lower separation section. By maintaining a constant value for the temperature of this stage, the bottoms composition hopefully remains essentially constant.

The considerations in selecting each control stage will be discussed shortly.

The following are associated with a temperature control stage:

Temperature measurement. A temperature probe is used to measure the temperature on the stage.

Temperature controller. The stage temperature is the controlled variable for a temperature controller. The manipulated variable for the temperature controller is usually a flow. For the upper control stage, the flow is either the distillate or the reflux. For the lower control stage, the flow is either the bottoms or the boilup (actually heat to reboiler). If possible, the output of each temperature controller is the set point for a flow controller, resulting in a temperature-to-flow cascade.

2.4.2. Target for Stage Temperature

The temperature profile from the design basis provides a value for the temperature on each control stage. Ideally, the set point for each temperature controller could be determined from the temperature profile. But in practice, both set points must be adjusted based on process operating conditions.

Figure 2.2 illustrates the role of the quality control laboratory. At some appropriate time interval, a technician collects a sample of the product stream and carries it to the QC lab for analysis. The product is deemed acceptable or unacceptable based on the results of this analysis. But in addition, the results of the analysis are communicated to the control room operator. Based on the results of the analysis, the control room operator may choose to adjust the set point for the respective stage temperature controller.

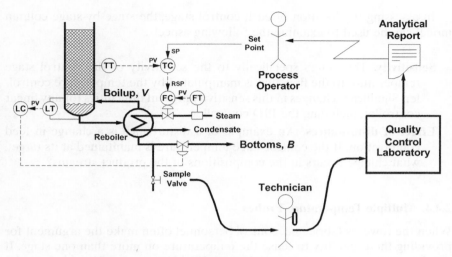

Figure 2.2. Role of QC lab.

In a sense, the control configuration in Figure 2.2 is a composition-to-temperature-to-flow cascade. PID controllers provide the flow control and temperature control. The process operator provides the composition control.

With time, the operators develop a "feel" for the effect of the control stage temperature on the analytical value (composition or physical property) of interest in the QC data. This "feel" is normally in the form of a sensitivity—a change of one degree in the control stage temperature will ultimately change the analytical value of interest by a certain amount. But in the end, it is the analytical value that must be controlled. Provided the analytical value is on target, the value of the control stage temperature is acceptable.

2.4.3. Location of Control Stage

The selection of the control stage involves two issues:

Dynamics. The control stage should be as near to the end of the tower as practical.

Sensitivity. Changes in the product composition must translate into a sufficiently large change in stage temperature that it can be used for control purposes.

The conventional logic for selecting the control stage focused on the change in temperature from one stage to the next. In the temperature profile in Figure 1.15, the temperature changes very little between stages 1 through 5. Consequently, the upper control stage should be stage 6, 7, or perhaps 8. The lower control stage should be stage 17, 18, or perhaps 19.

In assessing the location of each control stage, the stage-by-stage column model can be used to examine the following aspects:

Sensitivity. This refers specifically to the sensitivity of the control stage temperature to the flow that is manipulated by the temperature controller. Significant changes in this sensitivity present difficulties for any linear controller, including the PID controller.

Effect of disturbances. An example of a disturbance is a change in feed composition. If the control stage temperature is maintained at its target, what change occurs in the compositions of the product streams?

2.4.4. Multiple Temperature Probes

When the tower is fabricated, control personnel often make the argument for providing the capability to sense the temperature on more than one stage. If the design suggests that the control stage should be stage 6, then fabricate the tower with three temperature probes. In addition to the one at stage 6, install a second temperature probe at either stage 7 or possibly stage 8 and third temperature probe at either stage 5 or possibly stage 4. This is based on two concerns:

Design concerns. It is easy to become overconfident with the results of the stage-by-stage calculations. Caution should always be exercised for towers separating materials that deviate substantially from ideal behavior.

Change in service. Is the tower operating under the service for which it was designed? Control specialists can point to several situations where the answer is "no," usually due to factors that could not reasonably be foreseen at design time.

Adding temperature probes during tower fabrication is relatively inexpensive. But once the tower has been in service, adding temperature probes is always difficult and sometimes impractical (e.g., in towers with liners). However, getting project managers to spend the extra money is never easy.

2.4.5. Effect of Pressure on Control Stage Temperatures

For the depropanizer used as the example herein, the sensitivity of the control stage temperatures to column pressure can be determined by computing steady-state solutions for column pressures of 15.5 and 16.5 barg (± 0.5 barg relative to the column pressure of 16.0 barg for the base case). The following table presents the results for these two solutions and for the base case solution:

Column Pressure	Stage 6 Temperature	Stage 17 Temperature
15.5 barg	49.3°C	104.6°C
16.0 barg	50.4°C	106.1°C
16.5 barg	51.6°C	107.6°C
Sensitivity	2.3°C/barg	3.0°C/barg

For this tower, each 1-barg increase in pressure increases the stage 6 temperature by 2.3°C and the stage 17 temperature by 3.0°C.

For small excursions in column pressure, the behavior is essentially linear (the change in temperature for a +0.5 barg change in column pressure is approximately the same as the change in temperature for a −0.5 barg change in column pressure).

For larger changes in pressure, the nonlinear nature of the relationship would be reflected in the data. Since the pressure in most towers is controlled, large changes in pressure would not be expected and linear approximations to the relationship of temperature to pressure are satisfactory.

2.4.6. Pressure-Compensated Temperatures

The possibilities for stating the value of a gas flow are as follows:

Actual volumetric flow. This is the volumetric flow under flowing conditions (actual pressure and temperature).

Compensated volumetric flow. This is the volumetric flow under standard conditions (reference pressure and temperature).

The conversion from actual volumetric flow to compensated volumetric flow is based on an equation of state, the simplest being the ideal gas law.

For control stage temperatures, the counterpart is as follows:

Actual stage temperature. Stage temperature under tower operating conditions.

Pressure-compensated stage temperature. Stage temperature at a specified tower pressure.

Under ideal conditions, the pressure-compensated stage temperature would not be affected by changes in tower pressure. Linear approximations are normally used to make the conversion. Furthermore, the coefficients are affected by the composition on the stage. So in practice, changes in tower pressure will have some effect on the pressure-compensated stage temperature, but hopefully so small as to be insignificant.

Either or both control stage temperatures can be compensated for tower pressure. The temperature transmitter provides the actual control stage temperature; the pressure compensation relationship is applied to the actual control stage temperature to obtain the pressure compensated control stage temperature. The pressure-compensated control stage temperature is the measured variable for the respective control stage temperature controller.

Ideally, the pressure input to the pressure compensation relationship would be the pressure on the control stage. In practice, such pressure measurements are never available. Instead, the tower pressure is the pressure input to the pressure compensation relationship for both the upper control stage temperature and the lower control stage temperature.

Unfortunately, this introduces some error. The pressure at the control stage is higher than the tower pressure because of the pressure drop across the trays or packing between the control stage and the top of the tower. If this pressure drop is constant, the operators would compensate by adjusting the set point for the control stage temperature controller as per Figure 2.2. However, the pressure drop is a function of the vapor flow (approximately proportional to the square of the vapor flow). This introduces some variability to the pressure difference between the control stage and the top of the tower. The impact is larger on the lower control stage temperature than on the upper control stage temperature.

The pressure compensation can be applied in the opposite direction using the following approach:

- The measured variable for each control stage temperature controller is the actual control stage temperature.
- The operators specify the set point as the pressure-compensated control stage temperature, that is, the desired value for the control state temperature when the column pressure is at the reference value.
- The pressure compensation equation computes the desired value for the actual control stage temperature from the column pressure and the desired value for the pressure-compensated control stage temperature. Since the column pressure could change at any time, this computation must be repeated each time the PID is computed, not just when the operator changes the set point.

The selection of this approach vis-à-vis the approach originally proposed is a matter of preference; the performance is equivalent.

2.4.7. Practical Compensation Equation

For the depropanizer, previous calculations concluded that a 1-barg change in pressure caused the stage 6 temperature to change by 2.3°C, or a sensitivity of 2.3°C/barg. At least for small changes in the tower pressure, a linear approximation is satisfactory. Consequently, the equation for a pressure compensated temperature is as follows:

$$T_C = T - k(P - P_C),$$

where

T = actual stage temperature (°C);
P = actual tower pressure (barg);
T_C = pressure compensated stage temperature (°C);
P_C = pressure for pressure compensated stage temperature (barg);
k = sensitivity of stage temperature to pressure (°C/barg).

If the actual stage temperature is T and the actual tower pressure is P, then the stage temperature would be T_C if the tower pressure were P_C.

2.4.8. Theoretical Compensation Equation

The theoretical pressure compensation equation is based on the Clapeyron equation:

$$\frac{dT}{dP} = \frac{\Delta H_V}{R\,T\,(v_G - v_L)},$$

where

T = absolute temperature;
P = absolute pressure;
R = gas law constant;
ΔH_V = latent heat of vaporization;
v_G = specific volume of gas;
v_L = specific volume of liquid.

Assuming that the latent heat of vaporization ΔH_V is constant and that $v_G \gg v_L$ permits the Clapeyron equation to be integrated. When the column pressure is P, the stage temperature is T; when the column pressure is P_C, the stage temperature is T_C. The integrated equation is (all pressures and temperatures must be in absolute units)

$$\ln\frac{P}{P_C} = \frac{\Delta H_V}{R}\left[\frac{1}{T_C} - \frac{1}{T}\right].$$

Solving for T_C gives the following expression:

$$\frac{1}{T_C} = \frac{R}{\Delta H_V}\ln\frac{P}{P_C} + \frac{1}{T}.$$

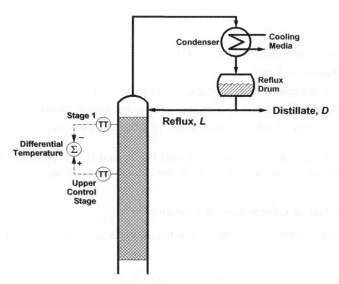

Figure 2.3. Differential temperature.

Such an equation could be implemented in digital controls. However, for the pressure variations encountered in towers, the far simpler linear compensation equation is quite adequate.

2.4.9. Differential Temperature

This approach is usually considered for towers where the product of interest is one component with relatively minor amounts of impurities. As illustrated in Figure 2.3 for the upper control stage, the temperature is measured on two stages:

- A stage near the end of the tower where the composition is largely one component. Changes in the impurities have little effect on this temperature; however, changes in pressure do affect this temperature.
- A stage within the separation section where the temperature is sensitive to changes in composition. This stage is normally the control stage. Of course, this temperature is also a function of pressure.

How does the pressure affect these two temperatures? If the effect were exactly the same on the two temperatures, the difference in the two temperatures would not be affected by pressure. In practice, the effect of pressure on the two temperatures is not exactly the same, so the temperature difference is affected by pressure. But for some towers, pressure affects the temperature difference much less than it affects the individual temperatures.

Before using the differential temperature configuration, the effect of pressure on the control stage temperature and the effect of pressure on the differential temperature should be determined from the stage-by-stage separation model. For the depropanizer being used as the example herein, the differential temperature will be computed from the stage 6 temperature (affected by composition) and stage 1 temperature (not affected by composition). The following table illustrates the effect of pressure on the stage temperatures and the differential temperature:

Column Pressure	Stage 6 Temperature	Stage 1 Temperature	Differential Temperature
15.5 barg	49.3°C	44.9°C	4.4°C
16.0 barg	50.4°C	46.3°C	4.1°C
16.5 barg	51.6°C	47.7°C	3.9°C
Sensitivity	2.3°C/barg	3.0°C/barg	−0.5°C/barg

Ignoring the sign of the sensitivity, the differential temperature approach reduces the sensitivity of the controlled variable from 2.3°C/barg for the stage 6 temperature to 0.5°C/barg for the differential temperature. The sensitivity for the differential temperature is significantly less, but the influence of pressure is not completely eliminated.

Issues arise pertaining to measuring the stage 1 temperature. If the reflux to the tower is significantly subcooled, a lower stage 1 temperature is a likely consequence. With air-cooled condensers, events such as afternoon showers in hot climates are a concern. Measuring the temperature on a lower stage in the tower must be considered, but the higher levels of impurities on the lower stages will also affect the temperature.

In some towers, differential temperature must be contemplated at the time the tower is constructed so that the temperature probe for the stage 1 temperature will be available. Later addition of temperature probes in lined towers, pressure towers, and so on, is difficult if even possible. In this respect, pressure compensation of the temperature measurements has a distinct advantage because it can be implemented with no additional measurements.

2.5. DISTILLATE COMPOSITION CONTROL: CONSTANT BOILUP

At this point, the focus is on configurations designed to control the composition of only one of the product streams. For some columns, this would be the distillate product; for other columns, this would be the bottoms product. Controlling the distillate composition is considered first; controlling the bottoms composition is the subject of a subsequent section.

When controlling only the distillate composition, one of the flows at the bottom of the tower can be specified directly. There are two options:

Constant boilup V (actually, constant heat input to the reboiler). The bottoms level must be controlled by manipulating the bottoms flow B.

Constant bottoms flow B. The bottoms level must be controlled by manipulating the boilup V, or in practice, by manipulating the heat input to the reboiler.

Of these, operating at constant boilup is more common and will be considered first.

The next chapter discusses the many condenser arrangements installed on towers. With a total condenser, the tower pressure is usually controlled by adjusting the heat removed in the condenser. To keep the diagrams simple, the pressure loop will not be included in the illustrations.

2.5.1. Constant Boilup

To maintain a constant boilup, a constant heat input is required. When the heating medium is steam (as will be the case in the illustrations), maintaining a constant steam flow provides a constant heat input to the reboiler. If the heating medium is a fluid such as hot oil, maintaining a constant fluid flow does not provide constant heat input. The rate of heat input must be calculated from measured values of the fluid flow, fluid inlet temperature, and fluid outlet temperature. The fluid flow must then be adjusted so that the rate of heat input is constant.

There are two possible control configurations for distillate composition:

Indirect material balance control (Fig. 2.4). The distillate composition is controlled by manipulating the reflux flow L; the reflux drum level is controlled by manipulating the distillate flow D.

Direct material balance control (Fig. 2.5). The distillate composition is controlled by manipulating the distillate flow D; the reflux drum level is controlled by manipulating the reflux flow L.

Table 1.3 listed the manipulated variables for a column in both the instrument context and the process context. Table 2.2 lists how each of these manipulated variables is used in the control configurations in Figures 2.4 and 2.5.

2.5.2. Relationship between Distillate Flow D and Reflux Flow L

The steady-state material balance around the condenser and reflux drum is written as follows:

$$V_C = L + D.$$

If the value of the overhead vapor flow V_C is known, the following are possible:

- Given a value of L, compute D as $D = V_C - L$.
- Given a value of D, compute L as $L = V_C - D$.

The simplest relationship is obtained by assuming that a constant heat input to the reboiler gives a constant boilup V and a constant overhead vapor flow V_C. The result is a linear relationship between L and D that can be established by the following observations:

Base point. At the base case, the distillate flow D is 22.8 mol/h and the reflux flow is 57.0 mol/h.

Slope. An increase of 1 mol/h in the distillate flow must translate to a decrease of 1 mol/h in the reflux flow.

The graph of this relationship is the dashed line in Figure 2.6.

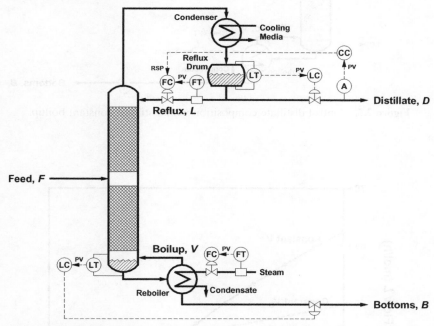

Figure 2.4. Control distillate composition with reflux: constant boilup.

TABLE 2.2. Control Configurations for Constant Boilup

Manipulated Variable	Indirect Material Balance Control (Fig. 2.4)	Direct Material Balance Control (Fig. 2.5)
Distillate flow D	Reflux drum level	Distillate composition
Bottoms flow B	Bottoms level	Bottoms level
Reflux flow L	Distillate composition	Reflux drum level
Condenser cooling	Column pressure	Column pressure
Heat input to reboiler	Constant boilup	Constant boilup

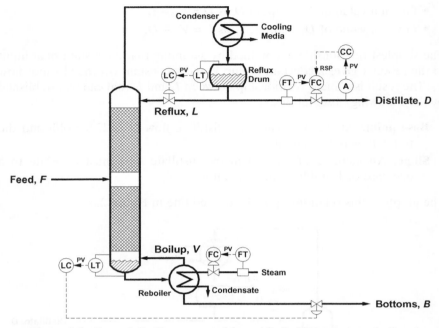

Figure 2.5. Control distillate composition with distillate: constant boilup.

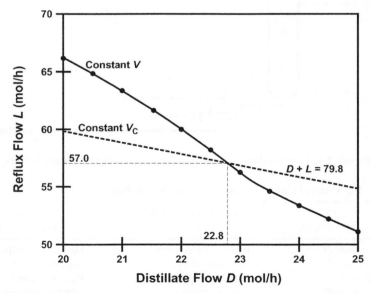

Figure 2.6. Relationship between reflux flow and distillate flow for a constant boilup.

More accurate plots of L as a function of D can be obtained using the stage-by-stage separation model. The solid line in Figure 2.6 is obtained by computing solutions for various values of the distillate flow D, but with the same boilup V as for the base case. A slightly different line would be obtained by computing solutions for various values of D, but with a constant heat input to the reboiler. The less ideal the materials being separated, the greater the differences between the graphs.

While the graphs differ depending on what assumptions are made (constant V_C, constant boilup, or constant heat input), all establish an algebraic relationship between L and D.

2.5.3. Manipulating D (Direct) versus Manipulating L (Indirect)

In the context of the steady-state column model, these two configurations are equivalent. The graph in Figure 2.6 establishes an algebraic relationship between L and D. In the steady-state calculations, the following two are equivalent:

• Specify L and obtain the value of D from the graph in Figure 2.6.
• Specify D and obtain the value of L from the graph in Figure 2.6.

Consider the two control configurations for distillate composition:

Manipulating reflux flow L (Fig. 2.4). The distillate composition controller specifies a value for the reflux flow L. At steady-state, the corresponding distillate flow D is determined by the graph in Figure 2.6.

Manipulating distillate flow D (Fig. 2.5). The distillate composition controller specifies a value for the distillate flow D. At steady-state, the corresponding reflux flow L is determined by the graph in Figure 2.6.

2.5.4. Separation versus Material Balance

At steady state, the compositions of the product streams are determined by two factors:

Separation. Separation is determined by the energy input, which for the control configurations in Figures 2.4 and 2.5, is determined by the boilup. This imposes a rather narrow range on the acceptable values of the reflux flow. Consequently, the separation is essentially fixed for both control configurations.

Material balance. This is determined by the distillate flow. For the control configuration in Figure 2.5, the distillate flow is specified directly. For the control configuration in Figure 2.4, the distillate flow is determined indirectly by the graph in Figure 2.6. These changes in D affect the product compositions through the column material balance.

To summarize, the separation is essentially fixed—boilup is essentially constant and only relatively small changes in the reflux flow are possible. The distillate composition controller influences the product compositions through the material balance either directly (by manipulating D as in Fig. 2.5) or indirectly (by manipulating L as in Fig. 2.4).

2.5.5. Operating Strategy

From an operations perspective, meeting the specifications for the distillate composition can be achieved with the following operating strategy:

1. Operate at a high or possibly maximum heat input to the reboiler. In many towers, this translates to operating just below the limits imposed by column flooding.
2. Rely on the distillate composition controller to adjust the column material balance until the desired distillate composition is attained.

Basically, the tower is always providing the maximum possible separation. The distillate composition controller attains the desired distillate composition by adjusting the distillate flow. But depending on the separation provided by the column, losses arise as follows:

Column is overdesigned with regard to separation. The composition of light key in the bottoms product is lower than required. The boilup is higher than necessary, which means more energy is being consumed than necessary.

Column provides inadequate separation. As the distillate composition controller reduces the distillate flow (either directly as in Fig. 2.5 or indirectly as in Fig. 2.4) to meet the distillate composition target, the amount of light key in the bottoms product increases. The consequences can be serious. For the depropanizer, the excess light key in the bottoms product makes it impossible for the butane product from the debutanizer to meet specifications.

The former consequence is the most common. In the days of cheap energy, towers were commonly operated in this manner. One argument for implementing double-end composition control is to avoid the excess energy consumption.

2.6. DISTILLATE COMPOSITION CONTROL: CONSTANT BOTTOMS FLOW

The control configuration in Figure 2.7 provides a constant bottoms flow. By specifying the set point to the bottoms flow controller, the split is fixed. The remaining loops in Figure 2.7 are as follows:

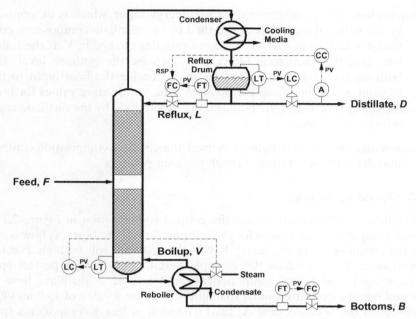

Figure 2.7. Control distillate composition with reflux: constant bottoms flow.

Bottoms level. Since the bottoms flow is constant, the level must be controlled by manipulating the heat input to the reboiler, which is equivalent to manipulating the boilup V.

Reflux drum level. Since the bottoms flow B is fixed, the value of the distillate flow D must be adjusted so as to close the column total material balance $(D = F - B)$.

Distillate composition. Since the distillate flow D is determined by the column total material balance, the manipulated variable for the distillate composition controller can only be the reflux flow L. In keeping with the practice within this book, Figure 2.7 includes a composition-to-flow cascade.

2.6.1. Separation versus Material Balance

For the control configuration in Figure 2.7, the composition of each product stream is determined by two factors:

Material balance. This is determined by the bottoms flow (or actually the B/F ratio). As will be discussed shortly, the set point for the bottoms flow is usually sufficiently high that some light key is present in the bottoms under all conditions.

Separation. This is determined by the energy input, which is determined by the value of the reflux L specified by the distillate composition controller. Changes in L lead to a corresponding change in V. If the reflux increases, the short-term effect is to increase the bottoms level. The bottoms level controller responds by increasing the heat input to the reboiler, which increases the boilup V. The steady-state values for both the reflux flow L and the boilup V are determined by the distillate composition controller.

To summarize, the material balance is fixed; the distillate composition controller manipulates the separation through the energy terms.

2.6.2. Operating Strategy

The difficulty with operating with the control configuration in Figure 2.7 is providing an appropriate value for the bottoms flow. If the bottoms flow is too low, the composition of the heavy key in the distillate will be high. For the depropanizer being used as the example herein (Fig. 1.14), a perfect split between light and heavy components would require a distillate flow of 23.4 mol/h and a bottoms flow of 76.6 mol/h. Suppose a value of 70.0 mol/h is specified for the bottoms flow. At least 6.6 mol/h of heavy components (primarily the heavy key) must be present in the distillate product.

From an operations perspective, meeting the specifications for the distillate composition can be achieved with the following operating strategy:

1. Operate with a conservatively high value for the bottoms flow.
2. Rely on the distillate composition controller to adjust the separation (through the energy terms) until the desired distillate composition is attained.

The consequence is that considerable light key will be present in the bottoms product—the higher the bottoms flow, the higher the composition of light key in the bottoms product.

2.6.3. Processes with Recycles

Operating in this manner would not be acceptable for the depropanizer. However, there are applications where this would be acceptable. A common process configuration is reaction followed by separation. Figure 2.8 illustrates a simple recycle process with two columns:

Column #1. The distillate is the desired product and must meet the specifications, including a limit on the amount of heavy key that may be present. The bottoms is the feed to the next column.

Column #2. The distillate is recycled to reactions. The bottoms is a waste stream.

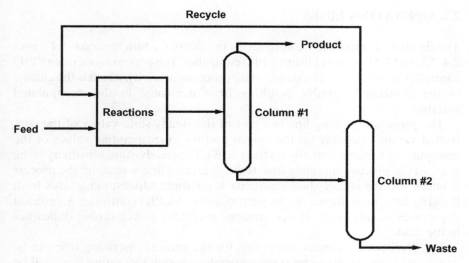

Figure 2.8. Simple recycle process.

Column #1 could be operated with a fixed bottoms flow, but a flow that is conservatively on the high side. This favors a low amount of heavy key in the distillate (the salable product), but at the expense of considerable light key in the bottoms. However, this light key is not lost. Almost all would be in the column #2 distillate, which is recycled back to reactions.

Although the light key for column #1 is not lost, some inefficiencies result:

- The light key in the feed to reactions effectively reduces the reactor volume available for reactions.
- Some light key might be lost through destructive reactions that consume the light key.

With experience, operations would establish a reasonable energy consumption (reflux flow, heat input to reboiler) in column #1. If the energy consumption is considerably less, the bottoms flow is too high (high bottoms flows make the specification for the heavy key in the distillate product easier to meet). If the energy consumption is considerably greater, the bottoms flow is too low.

While an attentive operations staff can lower the inefficiencies, use of double-end composition control on column #1 would lead to the most efficient process operation.

2.7. OPERATING LINES

The distillate composition controller in the control configurations in Figures 2.4, 2.5, and 2.7 is almost always a PID controller. The performance of any PID controller is affected by the steady-state process sensitivity, which is the change in the controlled variable resulting from a change in the manipulated variable.

The process operating line is a plot of the steady-state values of the controlled variable (usually on the y-axis) and the corresponding values of the manipulated variable (usually on the x-axis). The steady-state sensitivity is the slope of the process operating line. If the operating line is straight, the process is linear and the steady-state sensitivity is constant. Modest departures from linearity have little impact on the performance of a PID controller. Significant departures usually result in performance problems and/or tuning difficulties in the field.

For distillation columns, the points for the process operating line can be computed from the stage-by-stage separation model. Operating lines will be presented for

- the heavy key in the distillate (appropriate when composition control is used) and
- the upper control stage temperature (appropriate when temperature control is used).

2.7.1. Constant Boilup: Manipulating L

For the control configuration in Figure 2.4, Figure 2.9 presents the process operating lines. The manipulated variable for the process operating line is the reflux flow L. For the base case, the values are from Figure 1.14. For computing the operating line, the boilup is held constant at 64.88 mol/h.

The solution for the base case gives one point on each operating line. Additional points are computed by solving the stage-by-stage separation model for different values of the reflux flow L. For all solutions, the boilup is 64.88 mol/h. These points are then plotted to give the operating line in Figure 2.9.

The operating line for the heavy key in the distillate exhibits significant departures from linearity. For the feed composition in Figure 1.14, a perfect split between the light and heavy components would give a distillate flow of 23.4 mol/h, which translates to a reflux flow of approximately 56.4 mol/h. The operating line for the heavy key in the distillate exhibits a noticeable change in slope at a reflux flow of approximately this value. For lower reflux flows, the operating line exhibits a steep slope. For higher reflux flows, the slope is essentially zero, but primarily because the amount of the heavy key in the distillate is so small that its value is essentially zero when graphed as in Figure 2.9. The alternative of using a logarithmic scale for the composition will be examined shortly.

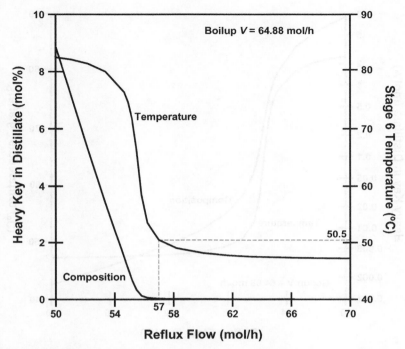

Figure 2.9. Distillate composition and stage 6 temperature as a function of reflux flow for control configuration in Figure 2.4.

The operating line for the upper control stage temperature also exhibits a significant departure from linearity. For reflux flows between 55 and 56 mol/h, the slope of the operating line is steep, which means that the process sensitivity is high. The upper stage temperature decreases by almost 20°C as the reflux flow changes from 55 to 56 mol/h. But for reflux flows above 58 mol/h and below 53 mol/h, the effect of reflux flow on upper control stage temperature is very low. The change in sensitivity will affect controller performance unless the column is operated within a narrow range of where the controller was tuned. These problems are likely to surface during large upsets to the tower.

Such extreme nonlinearities within a control loop usually translate to tuning problems in the field. A low sensitivity is needed for reflux flows between 55 and 56 mol/h. But if the low sensitivity is used for reflux flows outside this range, the controller responds too slowly. But if a higher sensitivity is used, stability issues arise for reflux flows between 55 and 56 mol/h.

2.7.2. Logarithmic Scale for Compositions

The graph for the composition in Figure 2.9 is unusable for very low amounts of the heavy key in the distillate. The operating line in Figure 2.10 uses a

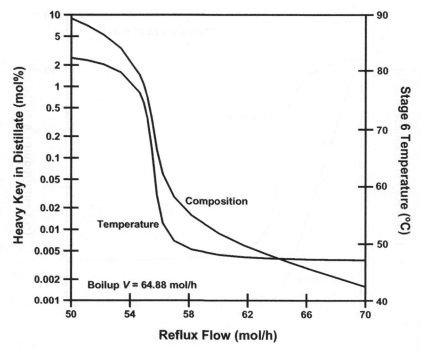

Figure 2.10. Log of distillate composition and stage 6 temperature as a function of reflux flow for control configuration in Figure 2.4.

logarithmic scale for the composition. The graph of the log of the heavy key in distillate (Fig. 2.10) is less nonlinear than the graph of the heavy key in distillate (Fig. 2.9). However, the logarithmic graph still exhibits significant nonlinearities, especially for reflux flows in the range of 56 mol/h.

For the operating line in Figure 2.10 to be appropriate to a control configuration, the controlled variable must be the log of the composition. This is sometimes applied in model predictive control configurations, the justification being that the relationship of composition to the manipulated variable (the operating line) is less nonlinear when the log of the composition is used.

With conventional controls, using log scales for the process variable (PV) input to a PID controller was impractical. Most digital controls provide a log function, so using the log of the composition for the PV input to a PID controller is an option. However, it does not seem to be common, probably because of the following:

• The linear scale in Figure 2.9 is essentially unusable when the composition of the heavy key is low. But from an operations standpoint, high values of an impurity are of most concern because the product may be unacceptable. Low values of an impurity rarely make the product unusable.

- Another argument is that the column is operated in a narrow range, so log scales are not necessary. However, upsets do occur, and this is when the log scale would be beneficial.

Using the log of the composition only applies to composition control configurations; it does not translate to temperature control configurations.

2.7.3. Constant Boilup: Manipulating *D*

For the control configuration in Figure 2.5, the manipulated variable for the operating line is the distillate flow *D*. This graph is presented in Figure 2.11 using a logarithmic scale for the composition. The points are computed by solving the steady-state separation model for different values of the distillate flow, but always with a boilup of 64.88 mol/h.

Each graph in Figure 2.11 is essentially a mirror image of the graph in Figure 2.10. The graphs are not exact mirror images due to the nonlinearities

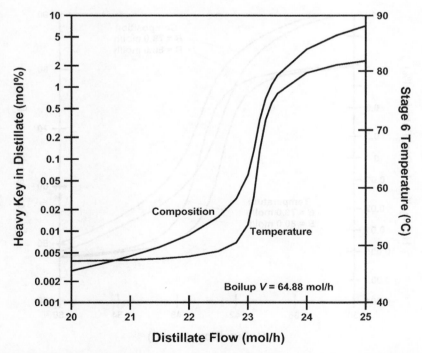

Figure 2.11. Log of distillate composition and stage 6 temperature as a function of distillate flow for control configuration in Figure 2.5.

in the plot of reflux flow as a function of distillate flow in Figure 2.6. But for the depropanizer, the nonlinearities are so small that their influence is not apparent from the graphs.

Because L and D are algebraically related, the observations previously stated for the graphs for the reflux flow can be translated to the graphs for the distillate flow. Consider the operating line for the upper control stage temperature. For distillate flows from 23.0 to 23.5 mol/min, the upper control stage temperature increases by approximately 20°C. Below a reflux flow of 22.5 mol/h and above a reflux flow of 24 mol/h, the sensitivity is much lower. The impact on an upper control stage temperature controller would be essentially the same.

2.7.4. Constant Bottoms Flow: Manipulating L

For the control configuration in Figure 2.7, the manipulated variable for the operating line is the reflux flow L. For a bottoms flow of 79.0 and 80.0 mol/h, the operating lines are presented in Figure 2.12 using logarithmic scales for the composition. The points are computed by solving the stage-by-stage sepa-

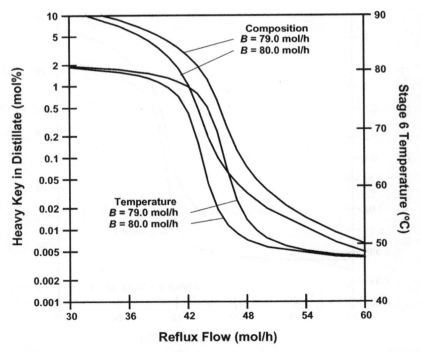

Figure 2.12. Log of distillate composition and stage 6 temperature as a function of distillate flow for control configuration in Figure 2.7.

ration model for different values of the reflux flow. For a given operating line, the points are always computed for the same bottoms flow.

The nonlinearities exhibited by the graphs in Figure 2.12 are comparable with the nonlinearities of those in Figures 2.10 and 2.11.

As noted previously, the operators are expected to occasionally change the bottoms flow. Increasing the bottoms flow decreases the heavy components in all stages, which is reflected in the following observations for Figure 2.12:

Composition. For a given reflux flow, the heavy key in the distillate is higher for a bottoms flow of 79.0 mol/h than for a bottoms flow of 80.0 mol/h. Increasing the bottoms flow basically shifts the operating line to the left.

Temperature. For a given reflux flow, the stage 6 temperature is higher for a bottoms flow of 79.0 mol/h than for a bottoms flow of 80.0 mol/h. Increasing the bottoms flow basically shifts the operating line to the left.

Changing the bottoms flow shifts the operating lines, but has little effect on the general shape and thus on the degree of nonlinearities to which the controller is exposed.

Approaches to address nonlinear behavior within the process include the following:

Characterization function. The operating line provides the basis for a characterization function that is usually inserted between the output of the composition or temperature controller and the set point to the reflux flow controller. Unfortunately, the fact that the operating line depends on the bottoms flow is a complication for this approach.

Scheduled tuning. Multiple sets of controller tuning parameters are provided. The set to be used is determined by the value of some process variable, which could be the manipulated variable (the reflux flow L), the controlled variable (heavy key in the distillate y_H or the stage 6 temperature T_6), or other variable (such as the bottoms flow B). The slopes of the operating lines in Figure 2.12 change at approximately the same values of the distillate composition and the stage 6 temperature, but not at the same values of the reflux flow L. Consequently, the following could be proposed:

Heavy Key in Distillate (mol/h)	Stage 6 Temperature (°C)	Controller Tuning
$y_H \leq 0.05$	$T_6 \leq 50$	Set 1, with a high controller gain
$0.05 < y_H < 3.0$	$50 < T_6 < 75$	Set 2, with a low controller gain
$y_H \geq 3.0$	$T_6 \geq 75$	Set 3, with a high controller gain

Tuning parameter sets 1 and 3 could possibly be the same.

2.8. TEMPERATURE PROFILES

The traditional approach to control stage selection is based on the change in temperature from one stage to the next. For the temperature profile presented in Figure 1.15, this suggests that the upper control stage should be stage 6, stage 7, or possibly stage 8.

Temperature profiles are only relevant when composition control is based on a control stage temperature instead of a direct composition measurement. For the purposes of this section, the distillate composition controller in Figures 2.4, 2.5, and 2.7 is replaced by an upper control stage temperature controller.

2.8.1. Effect of Manipulated Variable

The availability of a stage-by-stage separation model permits other issues to be taken into consideration in selecting the location of the control stage. The upper control stage temperature controller maintains the temperature at its set point by changing the controller output. As cascade is normally configured, the manipulated variable for the temperature controller is the set point for a flow controller.

The stage-by-stage separation model can determine the effect of changes in this flow on the temperature profile. That is, starting with the temperature profile for the base case, additional profiles are computed as follows:

- Increase the output of the temperature controller by some amount and compute the temperature profile.
- Decrease the output of the temperature controller by the same amount and compute the temperature profile.

Of particular interest is the amount by which the temperature profile changes, and if the change is the same for an increase and for a decrease in the controller output. This will be illustrated for the depropanizer. The temperature profile for the base case is the same as in Figure 1.15. For each configuration presented previously for controlling distillate composition, two additional temperature profiles are computed as follows (all flows are in mol/h):

Control Configuration	Constant Quantity	Manipulated Flow	Flow Change	Temperature Profiles
Figure 2.4	Boilup $V = 64.88$	Reflux L	$\Delta L = \pm 1.00$	Figure 2.13
Figure 2.5	Boilup $V = 64.88$	Distillate D	$\Delta D = \pm 1.00$	Figure 2.14
Figure 2.7	Bottoms $B = 77.20$	Reflux L	$\Delta L = \pm 3.00$	Figure 2.15

Table 2.3 presents numerical values for the flows and temperatures for the various solutions.

TABLE 2.3. Data from Temperature Profiles

(a) Manipulate reflux flow at constant boilup (Fig. 2.13)

	$\Delta L = -1.00$	Base Case	$\Delta L = +1.00$
Reflux flow, mol/h	56.00	57.00	58.00
Distillate flow, mol/h	23.04	22.80	22.55
ΔD, mol/h	+0.24	–	−0.24
Stage 6 temperature, °C	55.7	50.5	49.1
Stage 7 temperature, °C	62.8	54.9	52.3
Stage 8 temperature, °C	70.8	62.1	58.4

(b) Manipulate distillate flow at constant boilup (Fig. 2.14)

	$\Delta D = -1.00$	Base Case	$\Delta D = +1.00$
Distillate flow, mol/h	21.80	22.80	23.80
Reflux flow, mol/h	60.73	57.00	53.85
ΔL, mol/h	+3.73	–	−3.15
Stage 6 temperature, °C	47.9	50.5	79.1
Stage 7 temperature, °C	49.8	54.9	81.6
Stage 8 temperature, °C	54.0	62.1	83.3

(c) Manipulate reflux flow at constant bottoms flow (Fig. 2.15)

	$\Delta L = -3.00$	Base Case	$\Delta L = +3.00$
Reflux flow, mol/h	54.00	57.00	60.00
Boilup, mol/h	62.33	64.88	67.32
ΔV, mol/h	−2.55	–	+2.44
Stage 6 temperature, °C	54.7	50.5	48.9
Stage 7 temperature, °C	61.5	54.9	52.0
Stage 8 temperature, °C	69.5	62.1	57.8

Selecting the value for the flow change (ΔL and ΔD) involves the usual issues for distillation. The change must be sufficiently large for the effect to be discernable on the temperature profile graph. But if the change is too large, the nonlinear nature of distillation will distort the results. A reasonable starting point is 1% of the feed flow, which is 1.0 mol/h for a 100 mol/h feed flow. However, it may be necessary to use larger or smaller values.

2.8.2. Profiles for Constant Boilup

Figures 2.13 and 2.14 might suggest that the two control configurations affect the temperature profiles in a different manner, but actually their effect is the same. The differences in Figures 2.13 and 2.14 are due to the magnitude of the changes. From Table 2.3, the values of L and D for each profile are as follows:

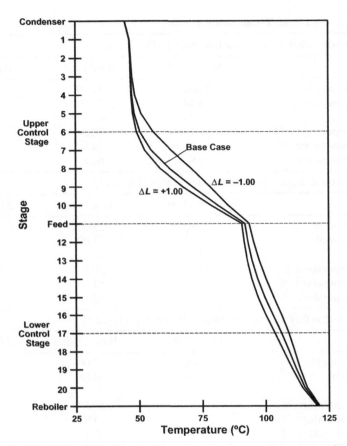

Figure 2.13. Effect of reflux flow on temperature profile: constant boilup.

Configuration	Change	Reflux Flow	Distillate Flow
Base case		57.00	$D = 22.80$
Figure 2.4	$\Delta L = -1.00$	56.00	$D = 23.04$ ($\Delta D = +0.24$)
Figure 2.4	$\Delta L = +1.00$	58.00	$D = 22.55$ ($\Delta D = -0.25$)
Figure 2.5	$\Delta D = -1.00$	$L = 60.73$ ($\Delta L = +3.73$)	$D = 21.80$
Figure 2.5	$\Delta D = +1.00$	$L = 53.85$ ($\Delta L = -3.15$)	$D = 23.80$

Assuming a constant overhead vapor flow V_C is obviously not good. Under this assumption, the change in reflux flow and the change in distillate flow should be equal but opposite (the dashed line in Fig. 2.6). For the above values, the changes are in the opposite direction, but are significantly different in magnitude. For $\Delta L = \pm 1.00$, the change in distillate flow is one-fourth of the change in reflux flow. For $\Delta D = \pm 1.00$, the change in reflux flow is over three times the change in distillate flow. However, the four points

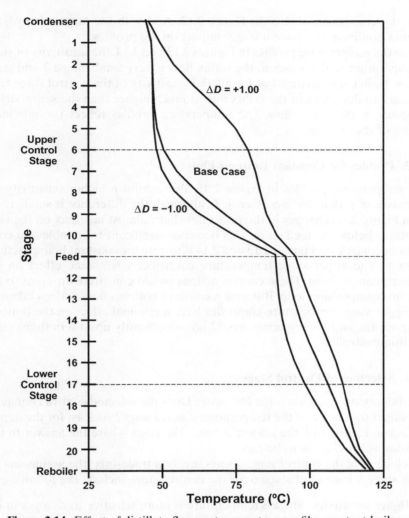

Figure 2.14. Effect of distillate flow on temperature profile: constant boilup.

$(D = 23.04, L = 56.00)$,

$(D = 22.55, L = 58.00)$,

$(D = 21.80, L = 60.73)$, and

$(D = 23.80, L = 53.85)$

all lie on the solid curve in Figure 2.6. The reflux flow and distillate flow are algebraically related according to the solid curve in Figure 2.6. The difference in the temperature profiles in Figures 2.13 and 2.14 are because the change $\Delta D = \pm 1.00$ in Figure 2.14 has a larger impact on conditions within the column than the change $\Delta L = \pm 1.00$ in Figure 2.13. And with the larger change, the column nonlinearities have a larger impact on the profiles.

For the temperature profiles in Figures 2.13 and 2.14, the sensitivity of stage 6 temperature to decreases in the reflux flow is very small. Stage 7 and stage 8 have higher sensitivities, but for all, the sensitivity of the control stage temperature to decreases in the reflux flow is much higher than the sensitivity to increases in the reflux flow. The temperature profiles reflect the nonlinear nature of the column.

2.8.3. Profiles for Constant Bottoms Flow

The temperature profiles in Figure 2.15 also exhibit a higher sensitivity for decreases in L than for increases in L, although the difference is smaller.

In Figure 2.15, changes in the reflux flow have almost no effect on the temperatures below the feed stage. This becomes significant for double-end composition control. As Figures 2.13 and 2.14 illustrate for constant boilup, actions taken by the upper stage temperature controller have some effect on the bottoms composition. These control actions would constitute an upset to the bottoms composition loop. But with a constant bottoms flow, actions taken by the upper stage temperature controller have a minimal effect on the bottoms composition, and consequently would not significantly upset a bottoms composition controller.

2.8.4. Selection of Control Stage

One interesting question is the following: Does the selected control configuration affect the choice of the temperature control stage? At least for the depropanizer in Figure 1.14, the answer is "no." The cases where the answer to this question is "yes" seem to be rare.

Selection of the control stage always involves trade-offs. The arguments for using stage 8 instead of stage 6 for the control stage include the following:

Higher sensitivity. Stage 8 temperature is more sensitive to changes in the manipulated variable than stage 6 temperature.

Less nonlinear. The sensitivity to increases in the manipulated variable is different from the sensitivity to decreases, but the difference is less for stage 8 than for stage 6.

The arguments against using stage 8 all stem from the fact that stage 8 is further from the distillate product stream than stage 6:

Slower dynamics. Composition dynamics are slow, so stage 6 temperature will respond more rapidly than stage 8 temperature.

Effect of off-key components. For the same distillate composition, changes in the off-key components will affect stage 8 temperature more than stage 6 temperature.

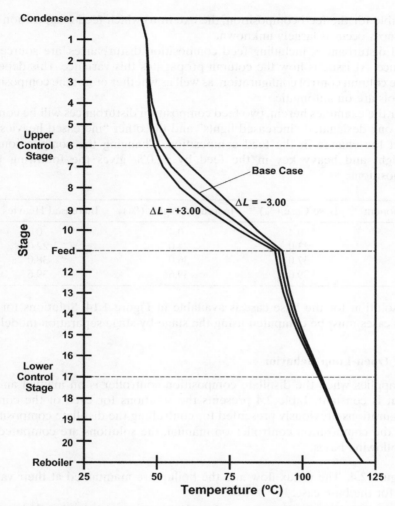

Figure 2.15. Effect of reflux flow on temperature profile: constant bottoms flow.

Selection of the temperature control stage is not an exact science. Incorporating multiple options for the temperature control stage into the fabrication of the column deserves serious consideration.

2.9. FEED COMPOSITION DISTURBANCES

Distillation columns are susceptible to a variety of disturbances, one possibility being changes in the feed composition. As on-stream analyzers are rarely available for the feed composition, the extent to which feed composition disturbances occur is largely unknown.

All disturbances, including feed composition disturbances, are sources of variance. At issue is how the column propagates this variance. This depends on the column control configuration, as well as whether or not the composition controls are on automatic.

For the examples herein, two feed composition disturbances will be considered, one designated "increased lights" and the other "increased heavies" to reflect the change in the feed composition. Changing the compositions of the light and heavy key in the feed by ±1.0% gives the following feed compositions:

Component	Base Case (%)	Increased Lights (%)	Increased Heavies (%)
C_2	0.4	0.4	0.4
C_3	23.0	24.0	22.0
C_4	37.0	36.0	38.0
C_5	39.6	39.6	39.6

The solution for the base case is available in Figure 1.14. Solutions for the other cases must be computed using the stage-by-stage separation model.

2.9.1. Open-Loop Behavior

This applies when the distillate composition controller is on manual and its output is constant. Table 2.4 presents the solutions for each of the control configurations previously presented for controlling the distillate composition. With the composition controller on manual, the solutions are computed on the following bases:

Figure 2.4. The reflux flow and the boilup are maintained at their values for the base case.

Figure 2.5. The distillate flow and the boilup are maintained at their values for the base case. Since the distillate flow is constant, the bottoms flow is also constant.

Figure 2.7. The bottoms flow and the reflux are maintained at their values for the base case. Since the bottoms flow is constant, the distillate flow is also constant.

The following observations apply to the solutions for Figures 2.5 and 2.7:

- The distillate and bottoms flows are the same.
- The values for the reflux flow and boilup are only slightly different.

Consequently, the differences between the two solutions are minor.

TABLE 2.4. Distillate Composition Controller on Manual (Open Loop Behavior)

Variable	Base Case	Increased Lights	Increased Heavies
(a) Manipulate reflux flow at constant boilup—Figure 2.4 (V = 64.88 mol/h; L = 57.00 mol/h)			
D	22.80 mol/h	23.72 mol/h	21.88 mol/h
L	57.00 mol/h	57.00 mol/h	57.00 mol/h
y_H	0.0283 mol%	0.0261 mol%	0.0315 mol%
T_6	50.4°C	50.1°C	51.0°C
T_8	62.1°C	60.8°C	63.7°C
B	77.20 mol/h	76.28 mol/h	78.12 mol/h
V	64.88 mol/h	64.88 mol/h	64.88 mol/h
x_{LK}	0.7803 mol%	0.8984 mol%	0.6733 mol%
(b) Manipulate distillate flow at constant boilup—Figure 2.5 (V = 64.88 mol/h; D = 22.80 mol/h)			
D	22.80 mol/h	22.80 mol/h	22.80 mol/h
L	57.00 mol/h	60.46 mol/h	54.15 mol/h
y_H	0.0283 mol%	0.0083 mol%	2.7503 mol%
T_6	50.4°C	48.0°C	80.0°C
T_8	62.1°C	54.1°C	84.1°C
B	77.20 mol/h	77.20 mol/h	77.20 mol/h
V	64.88 mol/h	64.88 mol/h	64.88 mol/h
x_{LK}	0.7803 mol%	2.0750 mol%	0.2941 mol%
(c) Manipulate reflux flow at constant bottoms flow—Figure 2.7 (B = 77.20 mol/h; L = 57.00 mol/h)			
D	22.80 mol/h	22.80 mol/h	22.80 mol/h
L	57.00 mol/h	57.00 mol/h	57.00 mol/h
y_H	0.0283 mol%	0.0136 mol%	2.4342 mol%
T_6	50.4°C	48.6°C	80.2°C
T_8	62.1°C	56.5°C	84.6°C
B	77.20 mol/h	77.20 mol/h	77.20 mol/h
V	64.88 mol/h	62.13 mol/h	67.36 mol/h
x_{LK}	0.7803 mol%	2.0766 mol%	0.2008 mol%

For the control configuration in Figure 2.5, feed composition changes are propagated largely to the product flows. A change in the feed composition of ±1.0% in the light key component means a change in the feed rate of this component by ±1.0 mol/h. This change is almost totally reflected in the change in the distillate flow and bottoms flow, resulting in only a small effect on the distillate composition (heavy key in the distillate y_H) and bottoms composition (light key in the bottoms x_L).

For the control configurations in Figures 2.5 and 2.7, feed composition changes are propagated largely to the product compositions. Comparing the total lights in the feed to the distillate flow suggests the following behavior:

Flows	Base Case (mol/h)	Increased Lights (mol/h)	Increased Heavies (mol/h)
Lights in feed	23.4	24.4	22.4
Distillate flow	22.8	22.8	22.8
Heavies in distillate	Small	Nil	0.4
Lights in bottoms	0.6	1.6	Small

The solutions in Table 2.4 give more precise values, but the conclusions are the same.

When both compositions are controlled manually, feed composition disturbances have the least effect on the product compositions when the operators adjust the reflux flow and the boilup. The level controllers determine the product flows by difference (indirect material balance). On this basis, it seems logical to use the following configurations for single-end composition control:

• Control distillate composition by adjusting the reflux.
• Control bottoms composition by adjusting the boilup.

In most cases, either will perform satisfactorily in a single-end control configuration. Then, it seems logical to use these configurations for double-end composition control. Unfortunately, this configuration usually has a high degree of interaction between the two composition loops, which creates problems in the field. The fact that

a. the distillate composition loop performs properly when used alone and
b. the bottoms composition loop performs properly when used alone

does not assure that they will perform properly when used together.

Suppose the desire is to use the control configuration in Figure 2.4, but with an upper stage temperature controller instead of the distillate composition controller. The change in T_6 is small ($-0.3°C$ for increased lights; $+0.6°C$ for increased heavies), but the change in the heavy key in the distillate is also small. As expected, using T_8 instead of T_6 gives a larger change ($-1.3°C$ for increased lights; $+1.6°C$ for increased heavies). But given the small change in the composition, using T_6 should be satisfactory.

2.9.2. Closed-Loop Behavior

This applies when the controller is on automatic and is maintaining a constant value for the distillate composition. As for the open-loop case, each of the control configurations presented previously for controlling the distillate composition will be considered. But with the distillate composition controller on automatic, the configurations in Figure 2.4 (manipulate reflux flow) and Figure 2.5 (manipulate distillate flow) are equivalent from a steady-state perspective.

That is, for a constant boilup, adjusting the reflux flow to achieve a specified distillate composition gives the same steady-state solution as manipulating the distillate flow to achieve the same distillate composition.

For this example, the upper control stage temperature will be controlled instead of the distillate composition. This permits the following question to be investigated: If the control stage temperature is controlled to a constant value, what variations occur in the distillate composition?

Using stage 6 for the upper control stage, Table 2.5a presents solutions for the base case, the increased lights, and the increased heavies. Maintaining T_6 at 50.5°C has the following effect on the composition of the heavy key in the distillate y_H:

TABLE 2.5. Upper Control Stage Temperature Controller on Automatic (Closed-Loop Behavior)

Variable	Base Case	Increased Lights	Increased Heavies
(a) Manipulate reflux flow or distillate flow at constant boilup—Figures 2.4 and 2.5 ($V = 64.88$ mol/h; $T_6 = 50.5$°C)			
D	22.80 mol/h	23.78 mol/h	21.83 mol/h
L	57.00 mol/h	56.75 mol/h	57.19 mol/h
y_H	0.0283 mol%	0.0303 mol%	0.0270 mol%
T_6	50.5°C	50.5°C	50.5°C
T_8	62.1°C	61.9°C	62.5°C
B	77.80 mol/h	76.22 mol/h	78.17 mol/h
V	64.88 mol/h	64.88 mol/h	64.88 mol/h
x_{LK}	0.7803 mol%	0.8207 mol%	0.7334 mol%
(b) Manipulate reflux flow or distillate flow at constant boilup—Figures 2.4 and 2.5 ($V = 64.88$ mol/h; $T_8 = 62.1$°C)			
D	22.80 mol/h	23.79 mol/h	21.82 mol/h
L	57.00 mol/h	56.72 mol/h	57.26 mol/h
y_H	0.0283 mol%	0.0310 mol%	0.0257 mol%
T_6	50.5°C	50.6°C	50.4°C
T_8	62.1°C	62.1°C	62.1°C
B	77.80 mol/h	76.21 mol/h	78.18 mol/h
V	64.88 mol/h	64.88 mol/h	64.88 mol/h
x_{LK}	0.7803 mol%	0.8099 mol%	0.7546 mol%
(c) Manipulate reflux flow at constant bottoms flow—Figure 2.7 ($B = 77.20$ mol/h; $T_6 = 50.5$°C)			
D	22.80 mol/h	22.80 mol/h	22.80 mol/h
L	57.00 mol/h	52.91 mol/h	200.00 mol/h
y_H	0.0283 mol%	0.0314 mol%	1.7573 mol%
T_6	50.5°C	50.5°C	91.1°C
T_8	62.1°C	61.8°C	96.1°C
B	77.20 mol/h	77.20 mol/h	77.20 mol/h
V	64.88 mol/h	58.88 mol/h	187.02 mol/h
x_{LK}	0.7803 mol%	2.0818 mol%	0.0009 mol%

Increased lights. y_H increases from 0.0283 to 0.0303 mol%.

Increased heavies. y_H decreases from 0.0283 to 0.0270 mol%.

The magnitude of the change in y_H is approximately the same as for the open-loop case in Table 2.4a, but the directionality is opposite. In both cases, the effect is relatively small.

Table 2.5b presents the same solutions but using stage 8 as the upper control stage. This gives a larger change in the distillate composition:

Increased lights. y_H increases from 0.0283 to 0.0310 mol%.

Increased heavies. y_H decreases from 0.0283 to 0.0257 mol%.

This supports the practice of using a control tray as close to the product stream as possible.

Table 2.5c presents the solutions for the control configuration in Figure 2.8 (constant bottoms flow; manipulate reflux flow to control distillate composition). For the increased lights case, the reflux can be reduced to attain a stage 6 temperature of 50.5°C. But for increased heavies, the reflux cannot be increased enough to attain a value of 50.5°C for T_6.

The solution in Table 2.5c is for the maximum possible reflux flow of 200 mol/h. The lights in the feed total 22.4 mol/h. With a distillate flow of 22.8 mol/h, the distillate must contain at least 0.4 mol/h (or 1.754 mol%) of a heavy component, primarily the heavy key. As discussed previously, this configuration can only be used when the value specified for the bottoms flow is sufficiently high that light components are present in the bottoms at all times.

2.10. BOTTOMS COMPOSITION CONTROL

For each configuration for controlling the distillate composition, there is a corresponding configuration for controlling the bottoms composition. The basic options are as follows:

Counterpart to Figures 2.4 and 2.5. Fix the reflux flow L and control the bottoms composition by manipulating either the boilup V (Fig. 2.16) or the bottoms flow B (Fig. 2.17).

Counterpart to Figure 2.7. Fix the distillate flow D and control the bottoms composition by manipulating the boilup V (Fig. 2.18).

2.10.1. Constant Reflux

For all control configurations presented herein, constant reflux is attained by providing a flow controller for the external reflux flow from a reflux drum.

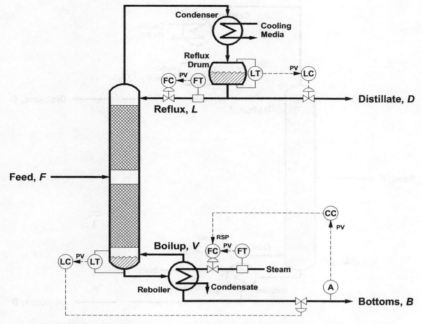

Figure 2.16. Control bottoms composition with boilup: constant reflux.

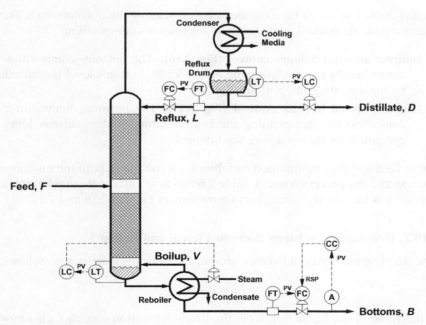

Figure 2.17. Control bottoms composition with bottoms flow: constant reflux.

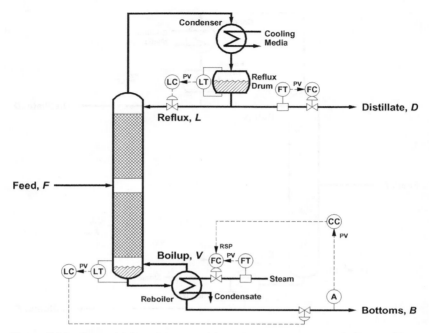

Figure 2.18. Control bottoms composition with boilup: constant distillate flow.

Reflux drum level must be controlled by manipulating the distillate flow. There are two possible control configurations for bottoms composition:

Indirect material balance control (Fig. 2.16). The bottoms composition is controlled by manipulating the boilup V; the bottoms level is controlled by manipulating the bottoms flow B.

Direct material balance control (Fig. 2.17). The bottoms composition is controlled by manipulating the bottoms flow B; the bottoms level is controlled by manipulating the boilup V.

Table 1.3 listed the manipulated variables for a column in both the instrument context and the process context. Table 2.6 lists how each of these manipulated variables is used in the control configurations in Figures 2.16 and 2.17.

2.10.2. Relationship between Bottoms Flow B and Boilup V

The steady-state material balance around the bottoms is written as follows:

$$L_B = V + B.$$

If the value of the liquid flow from the lower separation section L_B is known, the following are possible:

TABLE 2.6. Control Configurations for Constant Reflux

Manipulated Variable	Indirect Material Balance Control (Fig. 2.16)	Direct Material Balance Control (Fig. 2.17)
Distillate flow D	Reflux drum level	Reflux drum level
Bottoms flow B	Bottoms level	Bottoms composition
Reflux flow L	Constant	Constant
Condenser cooling	Column pressure	Column pressure
Heat input to reboiler	Bottoms composition	Bottoms level

- Given a value of V, compute B as $B = L_B - V$.
- Given a value of B, compute V as $V = L_B - B$.

The relationships between V and B for a constant reflux flow L are analogous to the relationships between L and D for a constant boilup V. Graphs analogous to those in Figure 2.6 for L and D can be developed for V and B. The simplest relationship is obtained by assuming that a constant reflux flow L gives a constant bottoms reflux L_B. However, a more accurate relationship can be developed using the stage-by-stage separation model. In either case, the boilup V and the bottoms flow B are algebraically related when the reflux flow is constant. Consequently, the configurations in Figures 2.4 and 2.5 are equivalent from a steady-state perspective.

2.10.3. Separation versus Material Balance

At steady state, the compositions of the product streams are determined by two factors:

Separation. Separation is determined by the energy input, which for the control configurations in Figures 2.16 and 2.17, is determined by the reflux flow. This imposes a rather narrow range on the acceptable values of the boilup. Consequently, the separation is essentially fixed for both control configurations.

Material balance. This is determined by the bottoms flow. For the control configuration in Figure 2.17, the bottoms flow is specified directly. For the control configuration in Figure 2.16, the bottoms flow is determined indirectly by closing the material balance around the tower bottoms. These changes in B affect the product compositions through the column material balance.

To summarize, the separation is essentially fixed—reflux flow is essentially constant and only relatively small changes in the boilup are possible. The bottoms composition controller influences the product compositions through the material balance either directly (by manipulating B as in Fig. 2.17) or indirectly (by manipulating V as in Fig. 2.16).

2.10.4. Operating Strategy

From an operations perspective, meeting the specifications for the bottoms composition can be achieved with the following operating strategy:

1. Operate at a high or possibly maximum reflux flow. The limit may be imposed by the condenser, but since the boilup changes along with the reflux, this may translate to operating just below the limits imposed by column flooding.
2. Rely on the bottoms composition controller to adjust the column material balance until the desired bottoms composition is attained.

Basically, the tower is always providing the maximum possible separation. The bottoms composition controller attains the desired bottoms composition by adjusting the bottoms flow. But depending on the separation provided by the column, losses arise as follows:

Column is overdesigned with regard to separation. The composition of heavy key in the distillate product is lower than required. The reflux flow is higher than necessary, which means more energy is being consumed than necessary.

Column provides inadequate separation. As the bottoms composition controller reduces the bottoms flow to meet the bottoms composition target, the amount of heavy key in the distillate product increases. The consequences are usually serious.

Excess energy consumption is usually the consequence. To avoid the excess energy consumption, double-end composition control must be implemented on the tower.

2.10.5. Manipulate Boilup at Constant Distillate Flow

The control configuration in Figure 2.18 provides a constant distillate flow. By specifying the set point to the distillate flow controller, the split is fixed. The remaining loops in Figure 2.18 are as follows:

Reflux drum level. Since the distillate flow D is fixed, the reflux drum level must be controlled by manipulating the reflux flow L.

Bottoms level. Since the distillate flow is constant, the value of the bottoms flow B must be adjusted so as to close the column total material balance ($B = F - D$).

Bottoms composition. Since the bottoms flow B is determined by the column total material balance, the manipulated variable for the bottoms composition controller can only be the boilup V (actually, the heat input to the reboiler). In keeping with the practice within this book, Figure 2.18 includes a composition-to-flow cascade.

2.10.6. Separation versus Material Balance

For the control configuration in Figure 2.18, the composition of each product stream is determined by two factors:

Material balance. This is determined by the distillate flow (or actually the D/F ratio). The distillate flow D is determined by the set point that the operator specifies for the distillate flow controller. If this set point is fixed, then the material balance is fixed.

Separation. This is determined by the energy input, which is determined by the value of the boilup V specified by the bottoms composition controller. Changes in the boilup V lead to changes in the reflux flow L. If the boilup increases, the immediate effect is to increase the reflux drum level. The reflux drum level controller responds by increasing the reflux flow L. In effect, the values of both the boilup V and the reflux flow L are determined by the bottoms composition controller.

To summarize, the material balance is fixed; the bottoms composition controller manipulates the separation through the energy terms.

2.10.7. Operating Strategy

The difficulty with operating with the control configuration in Figure 2.18 is providing an appropriate value for the distillate flow. If the distillate flow is too low, the composition of the light key in the bottoms will be high.

From an operations perspective, meeting the specifications for the bottoms composition can be achieved with the following operating strategy:

1. Operate with a conservatively high value for the distillate flow.
2. Rely on the bottoms composition controller to adjust the separation (through the energy terms) until the desired bottoms composition is attained.

The consequence is that considerable heavy key will be present in the distillate product—the higher the distillate flow, the higher the composition of heavy key in the distillate product.

Applications where this is acceptable include recycle processes similar to the one in Figure 2.8, but where the recycle stream is the distillate product and the bottoms product is the salable product for which meeting the composition specifications is crucial.

2.10.8. Operating Lines

Operating lines can be computed for the configurations for controlling the bottoms composition using approaches analogous to those for computing operating lines for the configurations for controlling the distillate composition:

Control Configuration	Operating Line
Figure 2.16	Plot of bottoms composition (or lower control stage temperature) as a function of boilup for a constant reflux flow
Figure 2.17	Plot of bottoms composition (or lower control stage temperature) as a function of bottoms flow for a constant reflux flow
Figure 2.18	Plot of bottoms composition (or lower control stage temperature) as a function of boilup for a constant distillate flow

A linear axis must be used for the lower control stage temperature, but a logarithmic axis is more appropriate for the bottoms composition. The degree of nonlinearity of these operating lines will be comparable with those for the control configurations for the distillate composition.

2.11. PROPAGATION OF VARIANCE IN LEVEL CONTROL CONFIGURATIONS

This section examines the effect of variance in boilup on other variables within the tower. The customary approach to maintain a constant boilup is to maintain a constant heat input to the reboiler. In practice, this minimizes but does not totally eliminate variations in the boilup.

When steam is the heating medium, installing a steam flow controller will reduce the variance in the steam flow to essentially zero. Flow controllers are very fast and will maintain the flow very close to its set point, even in the face of pressure upsets within the heating media supply system. However, variance in the following quantities results in variance in the boilup:

Enthalpy of the condensing steam. This varies with steam supply pressure, steam quality, and so on.

Latent heat of vaporization of the liquid inside the reboiler. This is a function of the bottoms composition.

2.11.1. Amplification versus Attenuation of Variance

The variance in the boilup will propagate to some extent to all other variables within the tower, including product flows and product compositions. When variance is propagated from one variable to another, there are two possibilities:

Amplification. The magnitude of the variance increases as it propagates from one variable to another. From a process operations perspective, this is usually undesirable. A minor problem in the original variable can become a major problem once it is propagated to another variable.

Attenuation. The magnitude of the variance decreases as it propagates from one variable to another. From a process operations perspective, this is usually desirable. Variance in variables such as boilup is a problem only should it propagate to an unacceptable degree of variance in a variable such as a product composition.

2.11.2. Impact of the Control Configuration

The degree to which variance is propagated depends on two factors:

Process characteristics. Once the process is designed and constructed, the manner in which the process itself contributes to the propagation of variance is usually fixed.

Control configuration. On a long-term basis, material balances, energy balances, and so on, must close. If variance is present in one variable in these relationships, closing the balance requires that variance be present in another variable in the relationship. The logic within the control configuration to close such balances affects how the variance is propagated.

From a steady-state perspective, the direct material balance and the indirect material balance control configurations for single-end composition control are equivalent. However, they are not equivalent when analyzed from the perspective of propagation of variance.

2.11.3. Issues Pertaining to Level Control

Control configurations are not immune to certain problems normally encountered in numerical analysis. Specifically, the following statement is usually presented very early in a course on numerical methods:

Never subtract two large numbers to obtain a small number.

Any error in either of the two large numbers will be amplified in the small number. A level control configuration can essentially be subtracting one large flow from another large flow to obtain a small flow. Variance in either large flow is amplified in the small flow.

In a total condenser, all overhead vapor is condensed. If the overhead vapor rate is V_C, the liquid flow from a total condenser will also be V_C, and this is the liquid input to the reflux drum.

The reflux drum level is maintained by manipulating either the input flow or one of the output flows. The preference is to control any level by manipulating the largest possible flow. For the reflux drum, the largest flow is the overhead vapor flow V_C. But in most configurations, the tower pressure controller manipulates the overhead vapor flow to control the tower pressure. If so, V_C cannot be manipulated by the level controller. This leaves two options:

- Control reflux drum level by manipulating the reflux flow L.
- Control reflux drum level by manipulating the distillate flow D.

The preference would be to control the reflux drum level by manipulating the larger of L and D. In most control configurations, one of the streams from the reflux drum is manipulated to control the distillate composition. The options are the following:

Indirect Material Balance Control (Fig. 2.4):

   Distillate composition. Control by manipulating the reflux flow. A reflux flow controller as the inner loop of a cascade configuration is normally recommended based on improved composition control.

   Reflux drum level. Control by manipulating the distillate flow. Including a distillate flow controller cannot normally be justified based on improved control of drum level.

Direct Material Balance Control (Fig. 2.5):

   Distillate composition. Control by manipulating the distillate flow. A distillate flow controller as the inner loop of a cascade configuration is normally recommended based on improved composition control.

   Reflux drum level. Control by manipulating the reflux flow. Including a reflux flow controller cannot normally be justified based on improved control of drum level.

As product composition control is far more crucial to plant operations than drum level control, composition control must take priority. However, the impact on drum level control cannot be totally ignored, especially in columns with either a very high or very low external reflux ratio.

2.11.4. Indirect Material Balance Control of Distillate Composition (Fig. 2.4)

The relationships are as follows:

   V_C = source of variance (variance in V propagates to variance in V_C);
   L = specified by the distillate composition controller;

$D = V_C - L;$
$B = F - D.$

Especially when a reflux flow controller is installed, the variance in the reflux flow L will be nil. Consequently, the variance in the overhead vapor flow V_C is entirely propagated to the distillate flow D.

The amplification of the variance depends on the external reflux ratio L/D. Consider a high reflux ratio ($L/D = 9$) and a low reflux ratio ($L/D = 1/9$). If the variance in the overhead vapor flow V_C is 1%, V_C can be expressed as 100 ± 1. The variance in the distillate flow D is as follows:

V_C	L/D	D	L	
100 ± 1	9	10 ± 1	90	Undesirable
100 ± 1	1/9	90 ± 1	10	OK

At low reflux ratios, the amplification of variance is modest. At high reflux ratios, the amplification is significant (for a reflux ratio of 9, the variance is amplified by a factor of 10). This suggests that the indirect material balance control configuration should be considered whenever the external reflux ratio L/D is less than unity. At very high reflux ratios, this configuration must be avoided.

The following observations are from the perspective of a drum level controller that is manipulating the distillate flow D:

External reflux ratio L/D > 1. The reflux flow L is larger than the distillate flow D. The reflux drum level controller is manipulating the smaller of the two flows, which is not the desirable situation.

External reflux ratio L/D < 1. The distillate flow D is larger than the reflux flow L. In this case, the reflux drum level controller is manipulating the larger of the two flows, which is the desirable situation.

The issues pertaining to reflux drum level control suggest that the indirect material balance control configuration is favored when the external reflux ratio L/D is less than 1.

2.11.5. Direct Material Balance Control of Distillate Composition (Fig. 2.5)

The relationships are as follows:

V_C = source of variance (variance in V propagates to variance in V_C);
D = specified by the distillate composition controller;
L = $V_C - D$;
B = $F - D.$

Since the distillate flow is maintained by a flow controller, the variance in the distillate flow D will be nil. Consequently, the variance in the overhead vapor flow V_C is entirely propagated to the reflux flow L.

As before, consider a high reflux ratio ($L/D = 9$) and a low reflux ratio ($L/D = 1/9$). For a variance in the overhead vapor flow V_C of 1%, the variance in the reflux flow L is as follows:

V_C	L/D	D	L	
100 ± 1	9	10	90 ± 1	OK
100 ± 1	1/9	90	10 ± 1	Undesirable

At high reflux ratios, the amplification of variance is modest. At low reflux ratios, the amplification is significant (for a reflux ratio of 1/9, the variance is amplified by a factor of 10). This suggests that the direct material balance control configuration should be considered whenever the external reflux ratio L/D is greater than unity. At very low reflux ratios, this configuration must be avoided.

The following observations are from the perspective of a drum level controller that is manipulating the reflux flow L:

External reflux ratio L/D > 1. The reflux flow L is larger than the distillate flow D. In this case, the reflux drum level controller is manipulating the larger of the two flows, which is the desirable situation.

External reflux ratio L/D < 1. The distillate flow D is larger than the reflux flow L. In this case, the reflux drum level controller is manipulating the smaller of the two flows, which is not the desirable situation.

The issues pertaining to reflux drum level control suggest that the direct material balance control configuration is favored when the external reflux ratio L/D is greater than 1.

2.12. LEVEL CONTROL IN DIRECT MATERIAL BALANCE CONFIGURATIONS

Figure 2.5 presented the direct material balance configuration for controlling the distillate composition. This section addresses an important issue that arises pertaining to the reflux drum level controller in the direct material balance control configuration. These do not arise for the indirect material balance control configuration.

To review, the control configuration in Figure 2.5 functions in the following manner:

- On an increase in the impurities in the distillate, the distillate composition controller reduces the set point to the distillate flow controller. The objective is to retain more of the light components in the tower, which will reduce the presence of the heavy key throughout the upper separation section.

- The reduction in the distillate flow causes the reflux drum level to increase.
- On an increase in the reflux drum level, the reflux drum level controller increases the reflux flow to the tower.

In this sequence of events, the reflux drum level controller translates the decrease in distillate flow to an increase in the reflux flow. The focus of this section is summarized by the following question:

What if the reflux drum level controller does not translate changes in the distillate flow to changes in the reflux flow in a timely manner?

2.12.1. Drum Level Controller Performance

For the direct material balance control configuration, the reflux drum level controller should be tuned to respond as quickly as possible. However, certain issues arise that complicate attaining a fast response in this level loop:

Large reflux drum. Large vessels respond slowly, and the level controller must be tuned accordingly.

Noisy drum level measurement. One approach to noisy measurements is to reduce the controller gain, which gives a slower response. The other approach is to provide filtering or smoothing. This adds lag to the loop, which will also require that the controller respond more slowly.

Drum level not a critical variable. Why maintain constant drum level? Swings in drum level can be tolerated as long as the low and high level switches do not initiate a shutdown. By following this logic, the drum level controller may be intentionally tuned to respond slowly.

In most control configurations, the impact of the above on the drum level control performance is tolerable. However, the direct material balance control configuration is an exception. A slow response in the drum level loop degrades the performance of the distillate composition controller.

2.12.2. Level Controller on Manual

This is the extreme case—changes in the distillate flow are not translated to changes in the reflux flow. Figure 2.19 presents the control configuration for distillate composition after removing the reflux drum level controller (the effective result when the controller is on manual).

This configuration functions in the following manner:

- On an increase in the impurities in the distillate, the composition controller reduces the distillate flow set point.

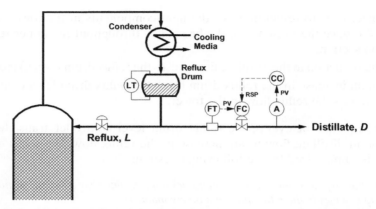

Figure 2.19. Direct material balance control of distillate composition with reflux drum level controller on manual.

- The reduction in the distillate flow causes the reflux drum level to increase.
- As is typical for integrating processes, the reflux drum level increases until the reflux drum high level switch initiates a shutdown.

With the reflux drum level controller on manual, the increase in distillate flow only causes the reflux drum level to increase. With no change in the reflux flow, the light components accumulate in the reflux drum and are not returned to the tower.

   With the reflux drum level controller on manual, the distillate composition controller cannot function. Changes in its output (the distillate flow) affect only the reflux drum level; they have no effect on the distillate composition. With the reflux drum level controller in manual, the composition controller is in open-loop automatic, which is an unstable configuration.

2.12.3. Dependence of Composition Loop on Drum Level Loop

In the direct material balance control configuration in Figure 2.5, the distillate composition loop can function only if the drum level loop is functioning properly. This dependence is very similar to the dependence of the outer loop of a cascade on the inner loop of a cascade. However, the arrangement of the composition loop and the level loop in Figure 2.5 is not a cascade configuration as the term is normally used in the industry.

   The term "cascade control" applies to configurations where the output of one controller is the set point to another controller. The control configuration in Figure 2.5 contains a composition-to-flow cascade. The output of the distillate composition controller is the set point to the distillate flow controller. The composition controller is the outer loop; the flow controller is the inner loop.

The composition controller is totally dependent on the flow controller, which leads to the following statements:

- The composition controller cannot function if the flow controller is in local, that is, not accepting the set point from the composition controller. Most digital control systems provide logic to suspend the composition control calculations when the flow controller (the inner loop) is not accepting the set point from the outer loop (the composition controller).
- A significant dynamic separation is required. The usual desire is for the inner loop (the flow loop) to be faster than the outer loop (the composition loop) by a factor of 5.

2.12.4. Relationship between Composition Controller and Level Controller

The focus is on the relationship between the distillate composition controller and the reflux drum level controller. By the customary use of the term, these are not cascade loops. However, they have two characteristics in common with cascade:

- If the reflux drum level controller is on manual, the distillate composition controller cannot function.
- The distillate composition controller is totally dependent on the reflux drum level controller.

Whenever one loop is totally dependent on another loop, a significant dynamic separation is required. Based on experience with cascade, the desire is that the dynamics of the reflux drum level loop be five times faster than the dynamics of the distillate composition loop. But due the factors cited previously regarding drum level controller performance, this dynamic separation is not assured.

2.12.5. Block Diagram

The block diagram in Figure 2.20 illustrates the relationship between the distillate composition controller and the reflux drum level controller:

- The output of the distillate composition controller is the distillate flow.
- The distillate flow affects the reflux drum level. There is no direct effect of the distillate flow on the distillate composition.
- The output of the reflux drum level controller is the reflux flow.
- The reflux flow affects both the reflux drum level and the distillate composition.

Note the presence of two loops in the structure in Figure 2.20. The "inner" loop contains the reflux drum level controller, and is totally within the "outer"

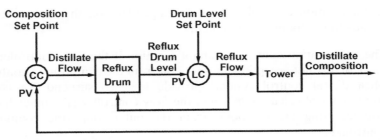

Figure 2.20. Block diagram of direct material balance control of distillate composition.

loop that contains the distillate composition controller. This "nesting" of loops is identical to that of cascade loops. The manner in which the two loops are connected is different. However, the key point is that whenever one loop is totally contained within another loop, a dynamic separation is required between the inner loop and the outer loop.

2.12.6. Attaining Required Dynamic Separation

There are two ways to attain the required dynamic separation between the distillate composition controller and the reflux drum level controller:

- Tune the reflux drum level controller to respond as rapidly as possible.
- Relax the tuning in the distillate composition controller.

The second practice slows the response of the distillate composition controller and degrades its performance.

The tower level controllers are always tuned before the composition controllers. This establishes the response speed of the reflux drum level loop. In order to obtain the required margin of stability, the distillate composition controller must be tuned so that its response is sufficiently slow to give the required dynamic separation. If the distillate composition controller responds too rapidly, cycling will occur in the distillate composition.

2.12.7. Alternate Configurations

Alternate configurations will be proposed to address the following problem: if the reflux drum level controller is on manual, the distillate composition controller will not function. An associated problem is that a slow reflux drum level controller seriously degrades the performance of the distillate composition controller. Fortunately, any configuration for which the distillate composition controller will function with the reflux drum level controller on manual

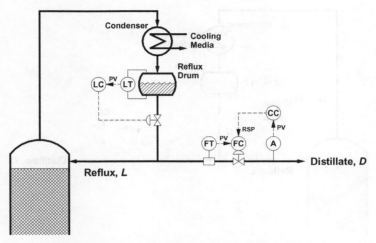

Figure 2.21. Control valve on common discharge line.

either eliminates or greatly reduces the impact of a slow reflux drum level controller on the performance of the distillate composition controller.

This provides a simple way to judge the viability of an alternate control configuration. If the distillate composition controller will function with the reflux drum level controller on manual, then the reflux drum level loop has little or no impact on the distillate composition loop.

2.12.8. Valve on Discharge Line from Reflux Drum

The advantage of this configuration is that it can be implemented with the same number of control valves and the same number of measurements. The configuration in Figure 2.21 is obtained by making only one change: the control valve previously in the reflux line is now in the common discharge line from the reflux drum. The flow through this control valve is the total flow from the reflux drum, which is $L + D$.

In the configuration in Figure 2.21, what is the effect of increasing the distillate flow? It will be some combination of the following:

- The total flow from the reflux drum will increase.
- The reflux flow to the tower will decrease.

How the effect is distributed depends on the pressure drops across the control valves.

Figure 2.21 may suggest gravity flow, but a pump is usually present on the reflux drum discharge. If most of the pressure drop is across the valve in the

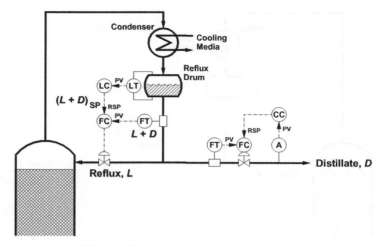

Figure 2.22. Measure total discharge flow.

reflux drum discharge line, a change in the distillate flow will be primarily translated into a change in the reflux flow. This is the desired result, and if this is indeed the primary consequence, then the distillate composition controller will function even if the reflux drum level controller is in manual.

2.12.9. Reflux Drum Discharge Flow Controller

The key to the configuration in Figure 2.22 is the flow transmitter in the common discharge line from the reflux drum. The flow transmitter senses the total discharge flow from the reflux drum, which is $L + D$. Using this measurement, a level-to-flow cascade is constructed:

- The output of the reflux drum level controller is the set point for the total discharge flow from the reflux drum. That is, this set point is the target for $L + D$.
- The output of the total discharge flow controller is the reflux control valve opening.

If there is a change in the distillate flow, the short-term effect is that the total discharge flow changes by this same amount. The total discharge flow controller then changes the reflux flow to bring the total flow back to its target. The net effect is that a change in the distillate flow is quickly translated to a change in the reflux flow. The dynamics are that of flow controllers.

The distillate composition controller will function provided the total discharge flow controller is in automatic. Switching the reflux drum level controller to manual has no effect on the distillate composition controller.

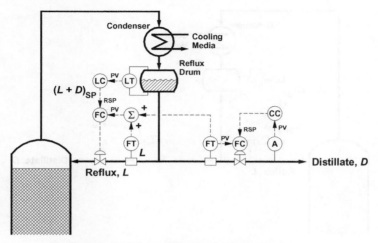

Figure 2.23. Compute total discharge flow.

2.12.10. Computed Reflux Drum Discharge Flow

Instead of directly measuring the reflux drum discharge flow, the control configuration in Figure 2.23 computes its value as follows:

- Measure the distillate flow D.
- Measure the reflux flow L.
- Sum these two measurements to obtain the reflux drum discharge flow.

The output of the summer is the measured variable for the reflux drum discharge flow controller. This is the inner loop of the level-to-flow cascade. If the distillate flow increases, the short-term result is an increase in the reflux drum discharge flow. The reflux drum flow controller changes the reflux valve opening to restore the reflux drum discharge flow to its set point. The net effect is that a change in the distillate flow is quickly translated to an equal and opposite change in the reflux flow. The dynamics are that of flow controllers.

The distillate composition controller will function provided the total discharge flow controller is in automatic. Switching the reflux drum level controller to manual has no effect on the distillate composition controller.

2.12.11. Computed Reflux Flow Set Point

A summer can be incorporated into the control configuration in two ways:

- Sum the distillate flow and the reflux flow to obtain the reflux drum discharge flow, which is then used as the measured variable for the reflux

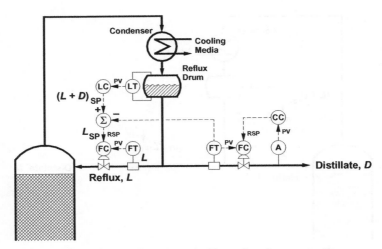

Figure 2.24. Compute set point for reflux flow controller.

drum discharge flow controller. This approach was the basis for Figure 2.23.

- Subtract the distillate flow from the target for the reflux drum discharge flow to obtain a target for the reflux flow. The schematic in Figure 2.24 reflects this approach.

The two approaches are basically equivalent, and the difference in performance will be trivial. Of the two, the configuration in Figure 2.24 is more often installed, probably because most are more comfortable with a reflux flow controller than a reflux drum discharge flow controller.

The distillate composition controller will function provided the reflux flow controller is in automatic and is using the set point computed by the summer. If so, switching the reflux drum level controller to manual has no effect on the distillate composition controller.

2.12.12. Bottoms Level

When the bottoms composition is controlled by manipulating the bottoms flow (direct material balance), issues arise that are analogous to those for reflux drum level. While the same control configurations could be proposed to rapidly translate a change in the bottoms flow to a change in boilup, there is one major difference—a measurement for the boilup V is never available. The impact is as follows:

Figures 2.21 and 2.22. There is no counterpart for the bottom of the tower.

Figures 2.23 and 2.24. These could be implemented based on an estimated value for the boilup.

When the heating medium is steam, a value for the units of boilup per unit of steam could be computed from the base case solution from the stage-by-stage separation model and the latent heat of vaporization for the steam. For heating media such as hot oil, the rate of heat transfer can be computed and then divided by the latent heat of vaporization for the steam to obtain an estimated value of the boilup.

While errors in the estimated value for the boilup are inevitable, using a counterpart to the configuration in either Figure 2.23 or 2.24 has the desired effect of quickly translating a change in the bottoms flow to a change in the boilup. These errors manifest as a disturbance to the bottoms level, to which the bottoms level controller must respond. Usually, the improvements in the bottoms composition control performance easily offset the issues that arise for bottoms level control.

3

PRESSURE CONTROL AND CONDENSERS

The chapter examines various approaches for controlling the pressure in a column, including the following:

- Total condensers. The pressure is normally controlled by adjusting the heat transfer in the condenser.
- Partial condensers. The pressure can potentially be controlled with the distillate flow.
- Atmospheric towers.
- Vacuum towers.

This chapter concludes with a discussion of pressure minimization.

In order to control column pressure, some mechanism is required to vary the heat transfer rate in the condenser. The possible mechanisms can be divided into two categories:

Media side. For water-cooled condensers, this means a valve on the cooling water. For air-cooled condensers, this means variable speed fans or louvers. Both control issues and process issues arise.

Process side. The most common approach is a flooded condenser, in which some condensate is retained within the condenser in order to reduce the effective surface area for heat transfer.

The number of permutations is surprisingly large.

Distillation Control: An Engineering Perspective, First Edition. Cecil L. Smith.
© 2012 John Wiley & Sons, Inc. Published 2012 by John Wiley & Sons, Inc.

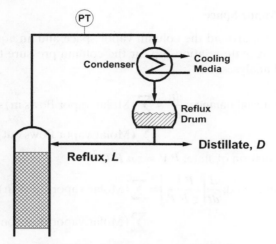

Figure 3.1. Column pressure measurement.

3.1. PRESSURE CONTROL

The usual practice is to measure the pressure at or near the top of the column. On piping and instrumentation (P&I) diagrams the pressure measurement is occasionally represented as physically connected to the column itself. But in practice, the column pressure measurement is usually physically connected to the overhead vapor line, as per the schematic in Figure 3.1.

The physical location of the condenser and the reflux drum may not be accurately represented by the P&I diagram either:

Small-diameter towers. An external structure is normally required for support. Physically locating the condenser and reflux drum at the top of the tower is customary. The column pressure transmitter will be physically near the top of the tower. Maintenance and other personnel can use the external structure to gain physical access to the transmitter.

Large-diameter towers. These towers are sufficiently rigid that an external structure is not required. But due to structural considerations, the condenser and reflux drum are physically located at or near grade level. The overhead vapor line comes down the tower to the condenser; a reflux pump is required to return the liquid to the top of the tower. For convenient physical access by maintenance and other personnel, the column pressure transmitter is often located in the overhead vapor line near its entrance to the condenser.

Although the term "column pressure" is routinely used (and will be used herein), the measured value is often the "condenser pressure."

3.1.1. Column Vapor Space

A material balance around the column vapor space and an equation of state provide an approximate relationship for the column pressure that is satisfactory for control analyses:

$$\text{Material balance: } \frac{dn}{dt} = \sum (\text{Molar vapor flows in}) - \sum (\text{Molar vapor flows out})$$

$$\text{Equation of state: } PV = z\,n\,R\,T$$

$$\text{Combined: } \frac{d}{dt}\left[\frac{PV}{z\,R\,T}\right] = \sum (\text{Molar vapor flows in}) - \sum (\text{Molar vapor flows out})$$

where

P = column pressure (absolute);
z = compressibility factor;
V = volume of column;
R = gas law constant;
n = moles of vapor in column vapor space;
T = column temperature (absolute).

3.1.2. Sources and Sinks for Vapor

The possible sources and sinks for vapor include the following:

Source or Sink	Remarks
Condenser	Usually a major sink for vapor
Reboiler	Usually a major source of vapor
Distillate	Sink for vapor if column has a partial condenser
Feed tray	Usually a small source or sink
Nonequimolal overflow	Usually a small source or sink
Vapor side steam	Sink for vapor (complex towers only)
Side cooler	Sink for vapor (complex towers only)
Side heater	Source for vapor (complex towers only)

The column pressure can only be effectively controlled by manipulating a major source or sink for vapor. This gives the following options for column pressure control:

Rate of condensation in condenser. The pressure in most towers is controlled in this manner.

Rate of vaporization in reboiler. Pressure can be controlled in this manner, but complications arise with regard to controlling the bottoms composition.

Distillate flow. This option is available only if the column is equipped with a partial condenser and the distillate flow rate is significant.

3.1.3. Heat Removal in Condenser

In most towers, the pressure is controlled through the heat removal in the condenser. The logic is as follows:

- On increasing pressure, the condensation rate in the condenser must be increased. This increases a major output term to the material balance around the vapor space.
- To increase the condensation rate, the rate of heat removal in the condenser must be increased.

At first glance, the simple approach is to install a control valve on the cooling media as illustrated in Figure 3.2. As will be explained in the next section, the effect of flow rate changes on the heat transfer rate decreases as the flow increases, approaching zero at high flows. The net result is that the configuration in Figure 3.2 often performs properly at low cooling media flow rates but not at high flow rates.

The complexity of the heat transfer relationships coupled with other issues that arise for column condensers has lead to a variety of condenser configurations for columns. Basically, this provides the subject matter for this chapter.

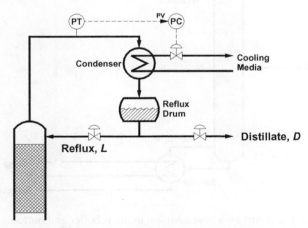

Figure 3.2. Pressure control via heat removal in the condenser.

3.1.4. Heat Addition in Reboiler

Although rarely used in practice, the column pressure can be controlled quite effectively through the reboiler. The logic is as follows:

- On increasing pressure, the vaporization rate in the reboiler must be reduced. This reduces a major input term to the material balance around the vapor space.
- To reduce the vaporization rate, the rate of heat addition in the reboiler must be reduced.

The most common heating medium is condensing steam, with the control valve on either the steam supply (as in Fig. 3.3) or on the condensate. Alternatives, such as hot oil, direct fired heaters, and so on, are most commonly encountered where the required reboiler temperature exceeds what can be achieved with steam.

The heat transfer relationships for the reboiler are analogous to those for the condenser and will be examined in the next chapter. But as observed previously, the column pressure is rarely controlled via the reboiler. The reasons

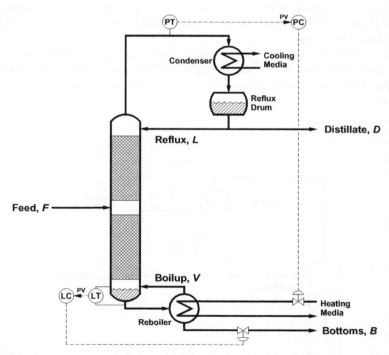

Figure 3.3. Pressure control via heat addition in the reboiler and bottoms level control via bottoms flow.

for this have nothing to do with the heat transfer relationships, but instead arise in the context of bottoms composition control.

3.1.5. Bottoms Level and Bottoms Composition

When the heating media control valve is manipulated to control the column pressure (as in Fig. 3.3), there is only one remaining control valve at the bottom of the tower: the control valve on the bottoms stream.

Additional control requirements at the bottom of the tower are as follows:

Bottoms level. Always required. When the heating media control valve is manipulated to control the column pressure, the bottoms level must be controlled by manipulating the bottoms flow, as illustrated in Figure 3.3.

Bottoms composition. Potentially required. But starting with the control configuration in Figure 3.3, what variable should be manipulated to control the bottoms composition?

When the column pressure is controlled via the heat addition to the reboiler, a problem arises with regard to controlling the bottoms composition. Is it feasible to control the bottoms composition by manipulating either the reflux flow or the distillate flow? The concern is that a manipulated variable above the feed stage is being manipulated to control a variable below the feed stage. The primary concern pertains to dynamics. The dynamic response of the bottoms composition to changes in boilup is far faster than its response to changes in either the reflux flow or the distillate flow.

3.1.6. Distillate Flow

When the column is equipped with a partial condenser, the distillate stream will be a vapor stream. In the schematic in Figure 3.4, the column pressure is controlled by manipulating the distillate flow.

In order for this approach to provide satisfactory pressure control, the distillate flow must be a significant term in the vapor space material balance. The parameter of interest is the ratio of the distillate flow to the condensation rate in the condenser. This ratio is determined by the external reflux ratio L/D. The total condensation rate is $L + D$. Therefore, the ratio of distillate flow to total condensation rate is

$$\frac{D}{L+D} = \frac{1}{L/D+1}.$$

If the external reflux ratio is 2, then the distillate flow D is 1/3 of the condensation rate in the condenser.

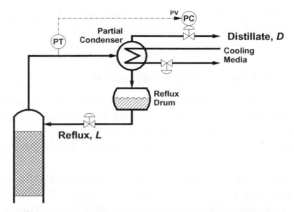

Figure 3.4. Pressure control via distillate vapor flow.

The smaller the distillate flow, the larger the swings in the distillate control valve opening that are required to control column pressure. If the distillate flow is too small, the swings will be so large that the control valve is frequently driven fully open and/or fully closed. When the control valve is fully open or fully closed, the column pressure is not being controlled.

What ratio of distillate flow to total condensation rate is required in order to be able to control the column pressure with the distillate flow? This depends on the disturbances to the tower—the larger the disturbances, the larger the required ratio of distillate flow to total condensation rate.

3.2. ONCE-THROUGH HEAT TRANSFER PROCESSES

For fluids such as cooling water and hot oil, there are two possible flow configurations:

Once-through—Figure 3.5a. The fluid makes a single pass through the heat transfer equipment. This is the simplest arrangement.

Recirculating—Figure 3.5b. A recirculation pump combines the fluid from the supply with some of the fluid from the exchanger exit to provide the fluid flowing into the exchanger. The recirculation rate is normally very high, giving a very small temperature rise from inlet to outlet and a uniform temperature differential.

On both condensers and reboilers, the once-through arrangement is most common. However, the recirculating configuration is occasionally installed in batch facilities that experience large variations in the heat transfer rates.

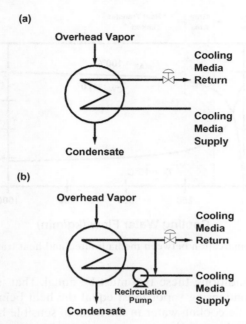

Figure 3.5. Heat transfer arrangements. (a) Once-through. (b) Recirculating.

In the once-through configuration, a low heat transfer means a low liquid flow and uneven temperature distribution within the heat transfer equipment. In the recirculating configuration, the liquid flow through the heat exchange equipment is high at all times, even when the flow from the supply is low.

The analysis that follows applies specifically to the once-through configuration.

3.2.1. Steps in Heat Removal

Consider removing heat in the condenser using cooling water. The heat that is removed from the condensing vapor is added to the cooling water that is exiting the exchanger. The heat removal proceeds as follows:

Transfer heat to cooling media. The usual heat transfer equations apply (heat transfer coefficient × area × mean temperature difference). This relationship imposes a maximum on the possible rate of heat removal by the condenser.

Remove heat by the cooling media. For cooling water, air, and so on, the heat is removed via sensible heat (flow × heat capacity × temperature rise).

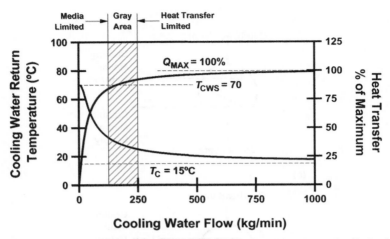

Figure 3.6. Demarcation between media limited and heat transfer limited.

At steady-state conditions, these two must be equal. That is, the heat transferred from the condensing vapor must equal the heat being removed from the exchanger by the cooling water in the form of sensible heat.

In air-cooled condensers, the steps are exactly the same. In reboilers with hot oil as the heating media, the steps are the same except that the direction of the heat flow is reversed.

3.2.2. Maximum Heat Transfer Rate

As illustrated in Figure 3.6, the cooling water exit temperature T_{CWR} is a function of cooling water flow W. At low cooling water flows, the exit temperature T_{CWR} is essentially the condensing vapor temperature T_C. As the cooling water flow W increases, the cooling water return temperature T_{CWR} decreases. At very high water flows, the temperature rise from supply to return is very small, and the return temperature T_{CWR} is essentially equal to the supply temperature T_{CWS}.

At very large cooling water flows, the following statements apply:

- The cooling water temperature rise from supply to return is very small.
- The return temperature T_{CWR} is essentially equal to the supply temperature T_{CWS}.
- The driving force for heat transfer is the condensing temperature T_C minus the cooling water supply temperature T_{CWS}.
- The heat transfer rate is the maximum possible and is determined solely by the heat transfer equations:

$$Q_{MAX} = U\ A\ (T_C - T_{CWS}),$$

where

Q_{MAX} = maximum heat transfer rate (kcal/h);
U = heat transfer coefficient (kcal/h m^2 °C);
A = heat transfer area (m^2);
T_C = temperature of the condensing vapor (°C);
T_{CWS} = cooling water supply temperature (°C).

Figure 3.6 presents the plot of the cooling water return temperature and the heat transfer rate as functions of cooling water flow. The ability to flow water through the condenser is oversized by approximately a factor of 4. As the cooling water flow rate increases from zero, the heat transfer rate initially increases with water flow. But the following statements apply at large cooling water flows:

- The cooling water return temperature approaches the cooling water supply temperature.
- The heat transfer rate approaches the maximum possible heat transfer rate.
- The sensitivity of the heat transfer rate to the cooling water flow approaches zero.

3.2.3. Heat Transfer Limited

At high cooling water flows, heat transfer from the condensing vapor to the cooling water is the limiting mechanism for heat transfer. The cooling water flow affects the heat transfer by altering the temperature difference for heat transfer. But at high cooling water flows, the temperature rise from cooling water supply to cooling water return is very small. If the cooling water return temperature is already close to its supply temperature, further increases in the cooling water flow will reduce the temperature rise from supply to return, but will have little effect on the temperature difference for heat transfer.

The numerical example in Table 3.1 may help clarify this point. Doubling the cooling water flow from 125 to 250 kg/min gives the following results:

TABLE 3.1. Effect of Cooling Water Flow on Heat Transfer Rate

W (kg/ min)	T_C (°C)	T_{CWS} (°C)	T_{CWR} (°C)	$T_{CWR} - T_{CWS}$ (°C)	ΔT_{LM} (°C)	Q (% of Q_{MAX})	$Q - Q_{MAX}$ (% of Q_{MAX})	$\dfrac{T_{CWR} - T_{CWS}}{T_C - T_{CWS}}$
125	70	15	33.1	18.1	45.3	82.4	17.6	0.33
250	70	15	25.0	10.0	49.8	90.5	9.5	0.18
500	70	15	20.2	5.2	52.3	95.1	4.9	0.10
∞	70	15	15	0.0	55.0	100.0	0.0	0.0

- Increases ΔT_{LM} by approximately 10%; consequently, the heat transfer increases about 10% (from 82.4% to 90.5% of max).
- Attains approximately 50% of the remaining possible heat transfer.

Doubling the cooling water flow from 250 to 500 kg/min gives the following results:

- Increases ΔT_{LM} by approximately 5%; consequently, the heat transfer increases about 5% (from 90.5 to 95.1 of max).
- Attains approximately 50% of the remaining possible heat transfer.

Doubling the cooling water flow again (to 1000 kg/min) would increase the heat transfer rate from approximately 95% of max to 97.5% of max. Each doubling of the flow attains approximately half of the remaining heat transfer. As the cooling water flow increases, its effect on the heat transfer decreases.

3.2.4. Onset of Heat Transfer Limited

At the onset of heat transfer limited behavior, there is a significant decrease in the slope of the graph of heat transfer as a function of cooling water flow. As the slope changes gradually, the "line in the sand" marking the onset of heat transfer limited is somewhat fuzzy, giving the "gray area" designated in Figure 3.6. For water flows less than about 125 kg/min, the process is not heat transfer limited. For water flows greater than 250 kg/min, the process is definitely heat transfer limited.

How can one detect the onset of heat transfer limited conditions? Consider using the cooling water temperatures. The maximum temperature difference for heat transfer is at the cooling water inlet, being the temperature of the condensing vapor less the cooling water supply temperature ($T_C - T_{CWS}$). The temperature increase from supply to return is $T_{CWR} - T_{CWS}$. The numerical example in Table 3.1 gives the value of the ratio of these two for various values of the cooling water flow. This suggests the following conditions for heat transfer limited:

$$\frac{T_{CWR} - T_{CWS}}{T_C - T_{CWS}} = \frac{\text{Temperature rise of the cooling water}}{\text{Maximum } \Delta T \text{ for heat transfer}} < 0.2.$$

However, the value of 0.2 is conservative; values of 0.25 or even 0.3 could be used.

3.2.5. Media Limited

At low cooling water flows, removal of heat by flowing cooling water through the condenser is the limiting mechanism for heat transfer. At low cooling water flows, all of the following are significant:

- The cooling water flow has a significant effect on the cooling water return temperature.
- The effect on the cooling water return temperature is sufficiently large that the temperature difference for heat transfer is significantly affected.
- The heat transfer rate changes to the same degree that the temperature difference for heat transfer is affected.

In order to control the condensation rate in the condenser using the valve on the cooling water, the exchanger must be operating in the media limited mode (changes in cooling water flow have a significant effect on the heat transfer rate). Once the transition is made to the heat transfer limited mode, the effect of changes in the cooling water flow on the heat transfer is too small to be usable to control the condensation rate in the condenser.

3.2.6. Excessive Cooling Water Flow

In properly designed heat transfer equipment, the following two capabilities should be comparable:

- The capability to pump cooling water through the heat transfer equipment.
- The capability to transfer heat from the process to the cooling water.

But for cooling media such as cooling water, the former capability often far exceeds the latter capability. When it comes to sizing piping, pumps, and other parts of the cooling water system, oversizing is common. But the oversizing is only in regard to the ability to pump cooling water through the condenser, which has little effect on heat transfer (and condensation rate) in the heat transfer limited region.

What is the consequence of oversizing? The cooling water valve operates at small openings. The behavior is similar to that of an oversized valve. At a low opening, the valve affects the condensation rate, but at a large opening, the valve has little effect on condensation rate.

3.3. WATER-COOLED CONDENSERS

Within this book, the "default" approach for varying the heat transfer rate in the condenser is through a control valve on the cooling media as in Figure 3.5a. This is the simplest to draw, but there are issues:

- Practical only when the cooling water has been conditioned. As will be explained shortly, this approach is not acceptable when the cooling water is natural water.

- The relationship between cooling water flow and the heat transfer rate is highly nonlinear. This is further complicated by the common practice of oversizing the ability to flow water through the condenser. In many installations, the cooling water valve normally operates at small openings. Once the valve is 50% open, further opening the valve has little effect on heat transfer. The valve has been properly sized for the capability to flow water through the condenser; it is the capability to flow water through the condenser that is oversized.

3.3.1. Control Valve Location

The cooling water control valve is usually located on the return. But if desired, it can be located on the supply. From the perspective of control performance, it makes no difference. The control valve affects the condensation rate in the condenser in exactly the same manner when located on the supply as when located on the return.

The decision is based entirely on other process considerations. The location of the control valve affects the pressure on the cooling water side of the condenser. The pressure on the water side is higher when the control valve is located on the return. In some applications, this difference in pressure affects the direction of any leaks in the exchanger (cooling water leaking into the process vs. process fluid leaking into the cooling water).

3.3.2. Btu Control

Btu control is sometimes considered to address the following problems:

- Btu control eliminates the nonlinear relationship between heat transfer and cooling media flow.
- Btu control responds to the disturbances in cooling media supply temperature.

The concept behind Btu control is simple:

- The heat transfer rate is computed from cooling water flow and temperature measurements.
- The computed heat transfer rate is the measured variable for a controller (often called a Btu controller) that manipulates the control valve on the cooling water.
- The column pressure controller adjusts the target for the Btu controller.

Some prefer an alternate configuration that computes the cooling water flow set point from the target for the heat transfer rate provided by the pressure controller.

3.3.3. Measurement Issues for Btu Control

The obvious problem for Btu control is the number of additional measurements: cooling water flow, cooling water supply temperature, and cooling water return temperature. However, the issues go beyond these:

- Btu control is often considered for applications where the heat transfer rate varies considerably, which is often the case in batch distillation. This means the cooling water flow will also vary considerably. This translates into a high turndown ratio requirement for the flow measurement, which means more expensive flow meters such as magnetic flow meters that are capable of a 50:1 turndown ratio.
- The Btu calculation entails computing $T_{CWR} - T_{CWS}$. Especially at high water flows, the difference will be small, which raises the numerical issues associated with subtracting two large numbers to obtain a small one. Instead of measuring the two temperatures, consider measuring the temperature difference directly. Smart temperature transmitters have the capability of sensing the temperature difference from two RTD inputs. This temperature difference is more accurate than the difference computed form the two temperature measurements.

In practice, Btu control is infrequently installed on condensers. Probably its main advantage is its capability to respond to disturbances in the cooling media supply temperature.

3.3.4. Issues Pertaining to Low Water Flows

Locating the control valve on the cooling media supply or return is a common practice in small chemical towers that use conditioned water for cooling. In large towers, the usual practice is to use natural water as the cooling water. The following issues arise at low water flow rates:

- Sediments in the natural water are deposited in the condenser, leading to premature fouling of the heat transfer surfaces.
- The higher cooling water return temperature decreases the solubility of the dissolved solids, causing precipitates to form on the heat transfer surfaces within the condenser.

These factors suggest that the flow of water through the condenser cannot be restricted. The high flows reduce the deposits and minimize the temperature rise between cooling water supply and cooling water return.

3.3.5. Varying the Heat Removal Rate

To control column pressure through the condenser, the control system requires some mechanism for varying the rate of heat transfer to the cooling

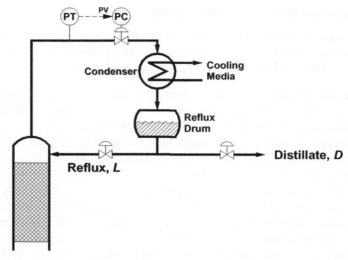

Figure 3.7. Pressure control using a control valve in the overhead vapor line.

water. Alternatives to a control valve on the cooling media include the following:

1. Valve in the overhead vapor line.
2. Flooded condenser.

The former is relatively simple and will be explained next. The flooded condenser will require a more lengthy discussion.

3.3.6. Valve in Overhead Vapor Line

In the configuration in Figure 3.7, the column pressure is controlled by manipulating a valve in the overhead vapor line. On rising pressure, the pressure controller should increase the opening of the valve, the results being as follows:

- The pressure in the condenser increases (approaches tower pressure).
- The condensing temperature increases.
- The temperature difference for heat transfer increases.
- The rate of heat removal and the condensation rate increase.

The effect of increasing the condensation rate is to reduce the tower pressure.

The column in Figure 3.7 is equipped with a total condenser—the overhead vapor flow is the distillate flow plus the reflux flow. For partial condensers with a sufficiently large distillate flow, the pressure is most likely controlled via a valve on the distillate vapor. But if the distillate flow is too small, column pressure cannot be controlled in this manner and the configuration in Figure 3.7 could be considered.

Designers are usually not in favor of anything that impairs condenser performance. During the summer season, many towers are limited by the cooling available in the condenser. Any additional pressure drop between the tower and the condenser reduces the maximum heat that can be removed in the condenser.

Two factors strongly favor installing a butterfly valve:

- The large size of the valve makes cost an issue. In large line sizes, butterfly valves are the least expensive.
- The valve inserts some pressure drop between the tower and the condenser. The pressure drop across a fully open butterfly valve is small, but not zero.

This approach is most frequently encountered in large towers operating above atmospheric pressure. It cannot be applied to the following:

- Towers with the condenser physically mounted on top of the column. These do not have an overhead vapor line.
- Vacuum towers. Further dropping the pressure in the condenser is impractical.

3.4. FLOODED CONDENSERS

The equation for heat transfer is as follows:

$$Q = U \, A \, \Delta T_{\mathrm{LM}}.$$

To affect Q, the control system must be able to influence one of the quantities on the right:

ΔT. Varying the cooling media flow affects ΔT provided the condenser is operating in the media limited region. A valve in the overhead vapor line affects the pressure in the condenser, which in turn affects ΔT.

U. Although not constant, the heat transfer coefficient does not change sufficiently to be used for control purposes.

A. The total heat transfer area is fixed by the design and construction of the exchanger. But by retaining condensate within the exchanger, the heat transfer area exposed to the condensing vapor can be any value smaller than the total heat transfer area. As will be explained shortly, the "A" in the heat transfer equation should be the exposed heat transfer

area. In condenser arrangements generally referred to as "flooded condensers," this term is a variable and can be used for control purposes.

In a condenser that is partially filled with condensate, some of the heat transfer
surface area is exposed to the condensing vapor and the remainder is submerged in the liquid condensate. The effects on heat transfer are as follows:

Exposed area. Heat transfer coefficients for condensing vapors are usually
very high, so the heat transfer area exposed to the condensing vapor will
provide most of the heat transfer.

Submerged area. The heat transfer area submerged in the condensate
largely serves to subcool the condensate. The heat transfer coefficients
are much lower, and the temperature differentials decrease with the
condensate temperatures. The submerged heat transfer surface area contributes very little to the total heat transfer.

Several mechanisms are available that enable a varying amount of liquid condensate to be retained in the exchanger. Herein the major approaches will be
examined; however, there are numerous variations.

3.4.1. Control Valve on the Condensate from the Condenser

A simple approach for retaining the condensate within the exchanger is to
install a control valve in the liquid line between the condenser and the reflux
drum, as illustrated in Figure 3.8. If the pressure in the tower is increasing, the
pressure controller must increase the opening of the condensate valve. This
drains more condensate from the condenser, exposing more heat transfer

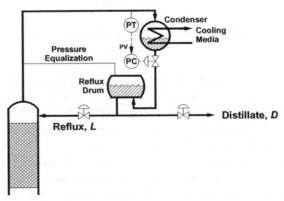

Figure 3.8. Control valve in the condensate line between the condenser and reflux
drum.

surface area and condensing more vapor. The condensate line from the condenser must enter the reflux drum at a point below the liquid surface.

Figure 3.8 also illustrates a line connecting the reflux drum vapor space to the overhead vapor line. Called a "pressure equalization" line, this is normally a small line and does not contain a control valve. However, the "pressure equalization" terminology is not quite accurate. The condensate from the flooded condenser is subcooled. Consequently, some hot vapors from the overhead line flow through the "pressure equalization" line and are condensed within the reflux drum. When a control valve is inserted into this line, it is normally called a "hot gas bypass." Such configurations will be discussed shortly.

Dynamically, the configuration in Figure 3.8 responds slowly. Opening the condensate valve has no immediate effect on the exposed heat transfer surface area and the condensation rate. As more condensate flows from the condenser, the level within the condenser slowly decreases to expose more heat transfer surface area. The effect on the tower pressure is also slow.

In the configuration in Figure 3.8, gravity provides the driving force for fluid flow (the condenser must be physically higher than the reflux drum). Control valve sizing requires values for the following:

Pressure drop across the control valve. This is the liquid head resulting from the difference between the liquid level in the condenser and the liquid level in the reflux drum.

Flow. The flow through the control valve is the total overhead flow (distillate flow plus reflux flow).

Good values are available for both.

3.4.2. Skin-Tight Reflux Drum

In the configuration in Figure 3.9, the reflux drum is completely filled with liquid (the pressure equalization line must be either removed or blocked off). This eliminates three items:

- Reflux drum level transmitter
- Condensate control valve
- Reflux drum level controller

The column pressure will be controlled by manipulating either the reflux flow or the distillate flow. The choice is actually dictated by how the distillate composition is controlled:

Distillate composition is controlled by manipulating the reflux. Tower pressure must be controlled by manipulating the distillate. This approach is the basis for the schematic in Figure 3.9.

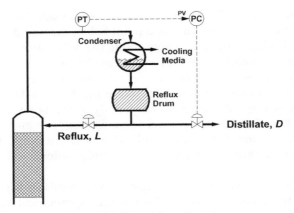

Figure 3.9. Skin-tight reflux drum.

Distillate composition is controlled by manipulating the distillate. Tower pressure must be controlled by manipulating the reflux.

If the reflux drum is completely full of liquid at all times, what purpose does it serve? From a control perspective, none at all. Removing the reflux drum has some appeal at the design stage. It saves some money and is also in keeping with the trend to reduce process inventory.

Before removing the reflux drum, some thought should be given to column startup. The reflux drum provides a reservoir of liquid that can be used as a source of reflux during startup. In most cases, either a reflux drum is required for startup or the design of the condenser must be modified to provide the necessary volume of liquid for column startup.

3.4.3. Hot Gas Bypass

The major drawback of the configurations in Figures 3.8 and 3.9 is the slow response of the column pressure to changes in the control valve opening. An alternative known as "hot gas bypass" is illustrated in Figure 3.10. The key aspects of this configuration are the following:

- The condenser is physically located at an elevation below that of the reflux drum. The liquid line from the condenser to the reflux drum enters below the liquid surface in the reflux drum.
- There is no valve in the overhead vapor line from the tower to the condenser. The pressure in the condenser is the same as the tower pressure.
- There is a line from the overhead vapor line to the reflux drum. This line is referred to as the hot gas bypass line. The control valve in this line is the hot gas bypass valve.

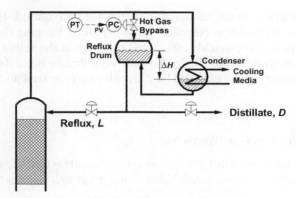

Figure 3.10. Hot gas bypass.

- The pressure in the reflux drum is less than the tower pressure. Opening the hot gas bypass valve increases the pressure in the reflux drum; closing the hot gas bypass valve decreases the pressure in the reflux drum.
- The difference between the pressure in the condenser and the pressure in the reflux drum determines the difference between the liquid surfaces in the reflux drum and in the condenser. The relationship is quite simple:

$$\Delta P_{HGP} = G\,\Delta H,$$

where

ΔP_{HGP} = pressure drop across hot gas bypass valve (cm H_2O);
ΔH = difference between liquid levels (cm);
G = specific gravity of the condensate.

For vertical exchangers, the exposed heat transfer surface area varies linearly with ΔH. For horizontal exchangers, the relationship depends on the geometry of the exchanger.

With respect to valve opening, the pressure controller in Figure 3.10 must be reverse acting. On increasing pressure, the controller should decrease the opening of the hot gas bypass valve, which gives the following results:

- The pressure in the reflux drum decreases.
- The difference between the level in the reflux drum and the level in the condenser increases.
- The level in the condenser decreases.
- The exposed heat transfer surface area increases.
- The heat transfer and the condensation rate in the condenser increase, which reduces the tower pressure.

The cause-and-effect relationships are somewhat convoluted. However, the hot gas bypass arrangement responds quite rapidly. Changing the opening of the hot gas bypass valve quickly affects the pressure in the reflux drum, which affects the hydrostatic head. The hydrostatic equilibrium is reestablished very quickly, which changes the exposed heat transfer surface area and the condensation rate.

3.4.4. Sizing the Hot Gas Bypass Valve

A common problem with hot gas bypass arrangements is that the control valve on the hot gas bypass is oversized. Valve sizing programs require the following information:

> **Pressure drop across the valve.** The physical locations of the condenser and reflux drum determine this pressure drop. The pressure drop across the valve (as height of water) is the difference between the level in the reflux drum and the level in the condenser, multiplied by the specific gravity of the condensate.
>
> **Flow through the valve.** All vapor flowing through the hot gas bypass valve is condensed in the reflux drum. The condensate is subcooled in the condenser, so the condensing vapor within the reflux drum provides the energy necessary to heat the condensate to the reflux drum temperature. Neither of these temperatures is precisely known. With no mixing in the reflux drum, stratification is very likely, so not all of the condensate is reheated to the reflux drum temperature. Unless data from a very similar operating tower is available, conservative values will be used, giving a flow that is too large and a hot gas bypass valve that is oversized.

3.4.5. Alternate Hot Gas Bypass Configuration

Numerous variations of hot gas bypass configurations are encountered. Herein only one more will be presented, specifically, the configuration in Figure 3.11. The configuration requires two control valves:

1. Control valve in the hot gas bypass line. The valve sizing difficulties are the same as for the configuration in Figure 3.10. But for reasons presented shortly, the consequences of valve oversizing on the configuration in Figure 3.11 are less than for the configuration in Figure 3.10.
2. Control valve on condensate flowing from the condenser to the reflux drum. The flow through this valve is the total condensate flow (distillate plus reflux), so the valve should be properly sized. And being on a liquid stream, the valve is smaller than valves inserted into the overhead vapor line (as in Fig. 3.7).

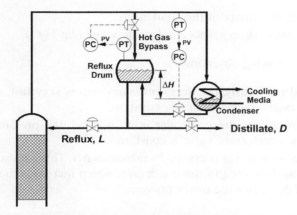

Figure 3.11. Alternate hot gas bypass.

The configuration in Figure 3.11 contains two pressure loops:

Reflux drum pressure. This pressure is controlled by manipulating the hot gas bypass valve. This is a very responsive loop (much faster than the tower pressure loop). This has two advantages:

1. From the perspective of the tower pressure loop, the reflux drum pressure is essentially constant.
2. Because the loop is very fast, the impact of an oversized hot gas bypass valve on the overall system performance is much less.

Tower pressure. This pressure is controlled by manipulating the valve in the condensate line. If the tower pressure is increasing, the tower pressure controller should increase the opening of the valve in the condensate line, which increases the exposed heat transfer area in the condenser.

One claim for the configuration in Figure 3.11 is that it provides better response to upsets to the tower pressure. That this is so is supported by the following equation that presents the relationship between the valve pressure drops and the difference in levels:

$$\Delta P_{HGP} = P_T - P_D = G\,\Delta H - \Delta P_C,$$

where

ΔP_{HGP} = pressure drop across hot gas bypass valve (cm H_2O);

P_T = tower pressure (cm H_2O);

P_D = reflux drum pressure (cm H_2O);

ΔH = difference between liquid levels (cm);

G = specific gravity of the condensate;

ΔP_C = pressure drop across condensate valve (cm H_2O).

Consider the following observations:

1. As noted previously, the drum pressure loop is very fast, so the reflux drum pressure P_D is essentially constant.
2. If the tower pressure controller is on manual, the pressure drop ΔP_C across the condensate valve is constant.
3. An increase in tower pressure P_T increases ΔH. This exposes more heat transfer surface area in the condenser, which increases the condensate rate and decreases the tower pressure.

Even without tower pressure control, the configuration in Figure 3.11 provides some compensation to disturbances in tower pressure. This reduces the correction that must be provided by the tower pressure controller, and hence improves the response to disturbances in tower pressure.

3.4.6. Separation of Set Points

The target for tower pressure never changes—at least that is often what they say. The reality is that changes in the tower pressure set point are small and infrequent, an example being operating at a lower tower pressure in winter than in summer. In the control configuration in Figure 3.11, the appropriate separation between the tower pressure and the drum pressure depends on design parameters and should not change. Consequently, a change in the tower pressure set point requires that the same change be made in the reflux drum pressure set point.

The simplest approach is to instruct the process operators to do this and assume that they will. The main issues are as follows:

1. Changes in the tower pressure set point are infrequent, so the burden on the operators is trivial.
2. The error rate for infrequently performed tasks is higher than the error rate for tasks that are routinely performed.

Clearly, the reflux drum pressure set point must be less than the tower pressure set point, but there is nothing in the control configuration in Figure 3.11 to assure that this is the case. This is a big red flag to those with the "anything that can go wrong will go wrong" and the "we never pass up the opportunity to make a mistake" perspectives.

Upon first inspection, a simple approach would be to change the controller for the hot gas bypass valve from a pressure controller to a differential pressure controller, the measured variable being the difference between the tower pressure and the drum pressure. However, this has an undesired consequence.

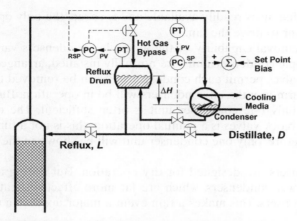

Figure 3.12. Maintaining proper separation of pressure set points.

The differential pressure controller would also be very fast, so any change in the tower pressure would be immediately reflected in the reflux drum pressure. With this change, the term $P_T - P_D$ in the equation

$$\Delta P_{HGP} = P_T - P_D = G \, \Delta H - \Delta P_C$$

presented previously is constant. If this term is constant, only a change in ΔP_C affects ΔH. With the reflux drum pressure controller as in Figure 3.11, a change in tower pressure is immediately translated to a change in ΔH, which is beneficial for responding to a tower pressure upset. Changing to a differential pressure controller eliminates this benefit.

Figure 3.12 presents a configuration that is easily implemented in digital control systems. The appropriate bias is subtracted from the tower pressure set point to provide the set point for the reflux drum pressure controller. Any change in the tower pressure set point is immediately reflected in the reflux drum pressure set point. The benefit of the improved response to pressure upsets is retained. Incorporating this addition into the control configuration is very easy for any digital control system. The major incentive is not to ease the workload on the operators, but to eliminate a potential source of errors during process operations.

3.5. AIR-COOLED CONDENSERS

Air-cooled condensers consist of horizontal, finned tubes over which air is directed upward by a fan that may be located either below or above the tubes. The horizontal footprint tends to be large, but the number of vertical rows of tubes is typically either two or four. The obvious appeal to this type of condenser is that air is readily available and is cheap, the only operating cost being the power to drive the fan.

The heat removal capability of the air-cooled condensers varies with the seasons. Multiple condenser units are normally installed, arranged in parallel paths. Block valves permit each condenser unit can be removed from service. During the warm months, all condensers will be in operation. But during the cool months, only one condenser unit is often sufficient. The opening and closing of the block valves is a manual operation; this is not a function of the controls. Therefore, only one condenser unit will be shown in the illustrations herein.

The condensers are designed for dry operation. But during a rain event, they become wet condensers, which are far more effective heat exchangers than dry condensers. This makes a rain event a major upset to a tower.

3.5.1. Manipulating the Airflow

Three options provide the capability to vary the airflow through the condenser:

Variable speed drive. In the schematic in Figure 3.13, the fan is equipped with a variable speed electric drive and a speed controller. Especially in older facilities, hydraulic or pneumatic drives are an alternative.

Louvers. Most air-cooled condensers are equipped with louvers that can be opened or closed manually. Equipping these with a pneumatic or electric actuator is feasible.

Variable pitch blades. These are the counterpart of the mechanisms used on airplane propellers.

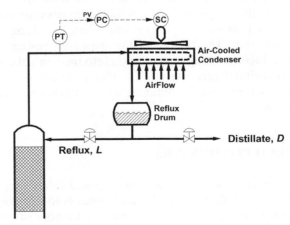

Figure 3.13. Varying the airflow through an air-cooled condenser.

Problems are often encountered when attempting to control variables such as tower pressure by varying the airflow. The usual experience is that reducing the airflow has no apparent effect on the heat transfer rate. There is a simple explanation for this. A water-cooled condenser was the basis for the previous discussion on media limited versus heat transfer limited modes of operation; these same concepts apply to air-cooled condensers.

The tendency is to oversize the capability to flow water through a condenser; the analogous practice is the norm with air-cooled condensers. Air is free; the only cost is the power to drive the fan. A common result is that the fan can blow far more air across the tubes than required. The air-cooled condenser is operating in the heat transfer limited mode, so changes in the airflow have almost no effect on the heat transfer. By the time the airflow is reduced sufficiently to be in the media limited mode, the fan speed is at its minimum or the louvers are almost closed. Either is the counterpart to operating a control valve barely off its seat.

Controlling tower pressure with airflow, via either variable speed drives or louvers, is definitely feasible. However, it requires a properly sized capability for the airflow. Usually, this means downsizing the existing capabilities. But in summer operations, the heat removal capability in the condenser is often the limiting factor for tower operations. This creates lots of skeptics for any proposal to reduce the airflow across an air-cooled condenser.

3.5.2. Valve in Overhead Vapor Line

Figure 3.7 presented a configuration for controlling the column pressure by manipulating a valve in the overhead vapor line. This is a simple and effective approach for air-cooled condensers as well. Closing this valve reduces the pressure within the condenser, which lowers the temperature within the condenser and the temperature differential for heat transfer.

The drawbacks for air-cooled condensers are the same as for water-cooled condensers:

- The large size of the valve makes cost an issue. Consequently, the valve is almost always a butterfly valve.
- The valve inserts some pressure drop between the tower and the condenser. During the summer season, towers with air-cooled condensers are usually limited by the cooling available in the condenser. Any additional pressure drop between the tower and the condenser reduces the maximum heat that can be removed in the condenser.

3.5.3. Flooded Condenser Arrangements

Air-cooled condensers are difficult to operate as flooded condensers. The vertical height is small (two or four rows of tubes vertically), and the tubes are usually large diameter. Small changes in the condensate level within the condenser have a large effect on the condensation rate. This has two consequences:

1. Proper control valve sizing is essential.
2. Nonidealities such as stiction and hysteresis in the control valve are amplified.

Figure 3.8 presented a configuration with the control valve on the condensate line from the condenser to the reflux drum. This configuration can also be applied to air-cooled condensers, the main disadvantage being the slow response of this configuration to upsets to the column pressure.

Figures 3.10 and 3.11 presented two versions of a hot gas bypass configuration for water-cooled condensers. These configurations (or variations thereof) are frequently installed on air-cooled condensers.

3.5.4. Weather Upsets

Process operators in refineries in arid locations such as west Texas can vividly explain the consequences of an afternoon rainstorm on a tower equipped with an air-cooled condenser. Even small amounts of moisture dramatically improve the heat transfer capability of the condenser. The reflux temperature drops rapidly (the degree of subcooling increases rapidly). The appropriate response is to decrease the external reflux flow accordingly. A control configuration known as "internal reflux control" provides this capability and will be presented in the subsequent chapter on feedforward control.

3.6. PARTIAL CONDENSERS

The choice of partial condenser versus total condenser is normally dictated by the nature of the distillate product. If the distillate product is largely a compound such as methane (such as the overhead product from a demethanizer), the considerations pertaining to the type of condenser are as follows:

Total condenser. The cooling media must be capable of attaining very low temperatures. Increasing the column pressure raises temperatures, but there are limits.

Partial condenser. The distillate product is a vapor; only the reflux must be condensed. The product from most demethanizers is natural gas, so there is no need to condense the methane. Most demethanizers have a partial condenser.

In a partial condenser, some of the overhead vapor is condensed as it flows through the condenser, giving a mixed vapor–liquid stream at the condenser exit. The vapor–liquid disengagement is provided by the reflux drum. Consequently, the vapor distillate product is withdrawn from the top of the reflux drum.

With a vapor distillate, controlling column pressure via the distillate stream is a potential approach. However, the primary objective of the column control configuration is to control the product composition(s). These considerations ultimately determine if the distillate composition is best controlled with the distillate or the reflux. The consequences on the column pressure control are as follows:

Control distillate composition with reflux flow (indirect material balance control). The distillate flow can be manipulated to control column pressure. However, if the distillate flow is very small relative to the boilup and condensation rates within the tower, effective control of column pressure cannot be realized using the distillate.

Control distillate composition with distillate flow (direct material balance control). Column pressure must be controlled in another manner, usually via the heat removal rate in the condenser.

In all control configurations presented in this section, the controls change the condensation rate in the condenser by manipulating a control valve on the cooling media. However, this is only because this configuration is the simplest to draw.

3.6.1. Pressure Control with Distillate Vapor Flow

Provided the distillate flow is sufficiently large, the control configuration in Figure 3.14 is viable. The arrangement of the control loops associated with a partial condenser is as follows:

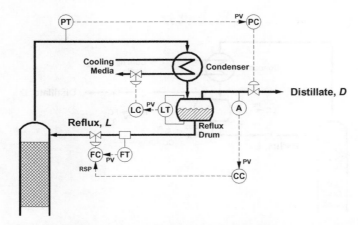

Figure 3.14. Pressure control by manipulating distillate vapor flow.

Distillate composition. Composition is controlled by manipulating the reflux flow using a composition-to-flow cascade.

Column pressure. The column pressure is controlled via the control valve on the distillate vapor flow.

Reflux drum level. The reflux drum level is controlled by manipulating the cooling to the condenser.

With the partial condenser arrangement in Figure 3.14, the condensation rate and the reflux rate must be equal at equilibrium conditions. The distillate composition controller determines the reflux rate. To maintain constant reflux drum level, the level controller must adjust the condensation rate until it is equal to the reflux flow.

As previously noted, the configuration in Figure 3.14 can only be applied to columns with a significant distillate flow, where "significant" is in the context of the overhead vapor flow. The smaller the distillate flow, the larger the changes (as a percentage of the distillate flow) required to compensate for upsets to the tower pressure. In the extreme cases, the pressure controller swings the distillate valve between fully closed and fully open, effectively resulting in on–off control and a cycle in the tower pressure.

3.6.2. Controlling Composition with Distillate Vapor Flow

As noted previously, composition control is critical and takes priority over all other control loops. If the distillate composition must be controlled using the distillate vapor flow, the result is the control configuration in Figure 3.15. The loops are as follows:

Distillate composition. This composition is controlled by manipulating the distillate flow via a composition-to-flow cascade.

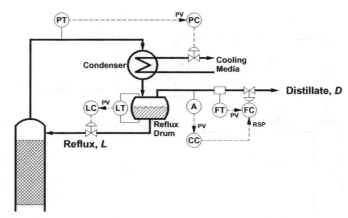

Figure 3.15. Composition control by manipulating distillate vapor flow.

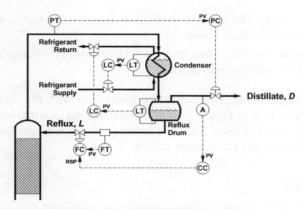

Figure 3.16. Refrigerant in the condenser.

Column pressure. The column pressure is controlled by manipulating the cooling to the condenser. This basically determines how much reflux will be returned to the tower.

Reflux drum level. The reflux drum level is controlled by manipulating the reflux flow.

The control configuration in Figure 3.15 is essentially the same as typically applied to total condensers.

3.6.3. Cooling with Refrigerant

Figure 3.16 presents the schematic of a column for which refrigerant is used as the cooling media. The overhead vapor flows through the condenser tubes; the vaporizing refrigerant is in the shell of the condenser. The refrigerant level controller maintains a constant refrigerant level in the condenser by manipulating the control valve installed on the refrigerant supply. The control valve on the refrigerant return affects the pressure on the refrigerant side of the condenser, which affects the temperature on the refrigerant side and the heat transfer rate. The reflux drum level controller adjusts the opening of this control valve so as to achieve a condensation rate in the condenser that is equal to the reflux flow rate specified by the distillate composition controller.

Composition control issues ultimately determine whether the distillate composition is controlled by manipulating the distillate vapor flow or by manipulating the reflux flow. In the configuration in Figure 3.16, the distillate composition is controlled by manipulating the reflux flow, the loops being as follows:

Distillate composition. The distillate composition is controlled by manipulating the reflux flow. The control configuration is a composition-to-flow cascade.

Column pressure. The column pressure is controlled by manipulating the valve on the distillate vapor.

Reflux drum level. The reflux drum level is controlled by manipulating the valve on the refrigerant return.

When the distillate composition must be controlled with the distillate vapor flow, the configuration in Figure 3.16 must be modified in a manner analogous to the configuration in Figure 3.15. The tower pressure is controlled by manipulating the valve on the refrigerant return.

3.6.4. Split Duty

A refrigerant is an expensive utility, especially as compared with cooling water. In some applications (one being deethanizers), the designers can significantly reduce the requirements for refrigerant by using a water-cooled (or air-cooled) partial condenser to provide the reflux and then use the refrigerant to condense the distillate product.

The configuration in Figure 3.17 contains two condensers:

Reflux condenser. The cooling media for this condenser is a low-cost utility such as cooling water or air.

Distillate condenser. Condensing the distillate requires low temperatures, which are achieved in Figure 3.17 using refrigerant.

The configuration in Figure 3.17 also contains two drums:

Reflux drum. The condensate from the reflux condenser is the reflux to the tower. The composition controller adjusts the reflux flow via a

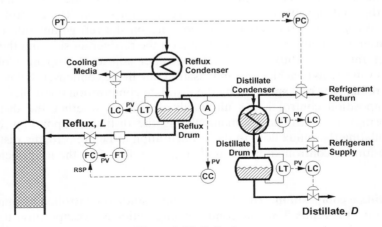

Figure 3.17. Split duty.

composition-to-flow cascade. The reflux drum level controller adjusts the cooling in the reflux condenser so that the amount of vapor condensed is the same as the reflux flow.

Distillate drum. The column pressure is controlled by manipulating the valve on the refrigerant supply to the condenser for the distillate product. This condensate becomes the distillate liquid product.

In the partial condenser configuration in Figure 3.16, the tower pressure is controlled by manipulating the control valve in the distillate vapor line leaving the reflux drum. There is no such valve in the split duty configuration in Figure 3.17. Inserting a control valve between the reflux drum and the distillate condenser would lower the pressure and consequently the temperature in the distillate condenser. The lower temperature would make it more difficult to remove the necessary heat in the distillate condenser, making some combination of the following necessary:

1. Operate the distillate condenser at a lower pressure on the refrigerant side, which means more work for the refrigerant compressor.
2. Increasing the heat transfer surface area in the distillate condenser.

Instead of manipulating a valve on the distillate vapor line between the reflux drum and the distillate condenser, the control configuration in Figure 3.17 controls the tower pressure by manipulating the control valve on the refrigerant return from the distillate condenser.

3.7. ATMOSPHERIC TOWERS

An atmospheric tower is operated either just above or just below atmospheric pressure:

Slightly above atmospheric pressure. The material inside the tower leaks to the outside, but no air leaks into the tower. This mode of operation is necessary when it is essential to prevent oxygen from entering the process.

Slightly below atmospheric pressure. Air leaks into the tower, but no material inside the tower leaks to the outside. This mode of operation is necessary when it is essential that none of the material within the tower leaks to the outside.

To minimize the leakage in either direction, the pressure set point should be as close to atmospheric as practical. However, it cannot be too close:

Towers operating above atmospheric pressure. A shutdown is initiated should the tower pressure fall below atmospheric.

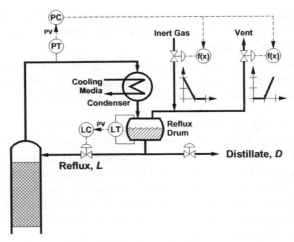

Figure 3.18. Atmospheric tower.

Towers operating below atmospheric pressure. A shutdown is initiated should the tower pressure rise above atmospheric.

The pressure measurement must be gauge pressure (the process pressure relative to atmospheric pressure) with a very narrow span.

3.7.1. Control Valves

As illustrated in Figure 3.18, the pressure is controlled using two control valves:

Vent control valve. Opening this valve releases gas to the vent. The gas is primarily noncondensables, but some product is always lost when gas is vented.

Inert gas control valve. Opening this valve admits an inert gas into the column. The inert gas may be anything that does not react with the material in the column. Nitrogen is certainly a possibility, but for some applications, the inert gas can be carbon dioxide, methane, or occasionally steam.

At an instant of time, the controller will be manipulating only one of these valves, with the other valve being fully closed. The logic for accomplishing this is normally called split range. The inert valve and the vent valve must never be open at the same time. This only releases the inert gas to the vent, which is not productive (but can be costly).

3.7.2. Characterization Functions

The schematic in Figure 3.18 implements the split range logic via two characterization functions, one for each control valve. Characterization functions are designated by "f(x)" and are sometimes referred to as function generators. An accompanying book [1] discusses the ideal and practical issues associated with split range control.

A major advantage of using characterization functions is that the split range adjustments are implemented entirely in the software. If these adjustments are implemented at the valve, then changes to the field equipment are required to make adjustments in the split range logic. One aspect of the split range logic for the atmospheric towers is that one of the control valves must begin to open slightly above mid-range, and the other control valve must begin to open slightly below mid-range. At small openings, most control valves will deviate from ideal behavior. The only way to accommodate this type of behavior is to "tune" the split range logic to the behavior exhibited by the control valves.

3.8. VACUUM TOWERS

In most vacuum towers, the condenser is physically mounted on the top of the tower. Most of the condensable vapors are condensed in the condenser, so the flow from the top of the condenser is largely the noncondensables.

The concern with overhead vapor lines on vacuum towers is the pressure drop. In vacuum service, vapor density is very low, which leads to high vapor velocities. The associated pressure drop can be very significant, especially when considered in the context of the total pressure. These considerations must be balanced against the structural issues associated with physically locating the condenser on top of the tower. The higher the vacuum, the greater the incentive to locate the condenser on top of the tower.

In the configuration in Figure 3.19, all of the condensate is removed from the tower to an external reflux drum. The chimney tray permits all vapor to pass to the condenser. All condensate is captured on the chimney tray and flows by gravity to the reflux drum.

The considerations pertaining to the external reflux drum are as follows:

- There are various mechanisms for withdrawing only the distillate product from the tower. With a slight modification to the chimney tray, the distillate product can be withdrawn, with the excess condensate (above the distillate flow) overflowing a weir to provide the reflux. This eliminates the reflux drum, the reflux drum level transmitter, the reflux drum level controller, and the reflux valve. These cost savings appeal to designers, so external reflux drums are the exception. However, there is no way to measure the reflux flow.

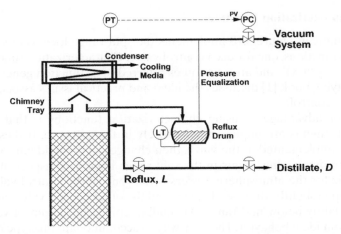

Figure 3.19. Control valve between the tower and vacuum system.

- When the separation section is packing, the reflux flow must always be above the flow required to wet the packing. With the external reflux drum, a flow measurement and flow controller can be provided for the reflux flow. The minimum flow required to wet the packing can be enforced through a minimum allowable set point for this flow controller.

Cost considerations at the design stage generally prevail, which means that most towers do not have an external reflux drum.

3.8.1. Valve on Vacuum Line

Figure 3.19 presents a configuration that controls the vacuum by manipulating a control valve installed in the piping that connects the tower to the vacuum system. Two relationships pertain to the pressure drop ΔP_V across the control valve:

- The pressure drop ΔP_V is the difference between the tower pressure P and the pressure at the vacuum source P_V.
- The pressure drop ΔP_V depends on the valve opening M and the flow F through the control valve.

These yield the following relationship:

$$\Delta P_V = P - P_V = f(M, F).$$

For a given flow through the control valve, the pressure controller must adjust the valve opening M until the desired pressure drop is attained. The relationship $f(M, F)$ depends on the valve size, valve characteristics, and so on. The smaller the flow F, the smaller the valve opening M required to give the desired ΔP_V.

Now for the problem. What if the flow F is zero? There is no pressure drop at any valve opening. The flow will never be exactly zero, but it can be so small that the valve cannot position to the required small opening. The result is a cycle in the pressure caused by the valve cycling between fully closed and the minimum sustainable valve opening. Very low flows are most commonly encountered in batch production facilities. In these facilities (and in any other application where the flow through the control valve is at times essentially zero), the configuration in Figure 3.19 is unacceptable.

3.8.2. Constant Inert Gas Bleed

There is a simple approach to ensuring that the flow through the vacuum control valve is not too low: Bleed an inert gas into the system upstream of the vacuum control valve. Usually, only a small line is required with the flow rate for the inert gas adjusted via a needle valve in the inert gas line. If the vacuum control valve is operating at too low an opening, the operators need only to increase the opening of the needle valve on the inert gas.

The term "inert gas" does not necessarily mean nitrogen. Other possible inert gasses include carbon dioxide, methane, or occasionally steam. Any gas that does not affect the process is acceptable, and wherever possible, a lower cost inert gas is used.

3.8.3. Pressure Control with Inert Gas Bleed

The schematic in Figure 3.20 controls the tower pressure by manipulating a control valve on the inert gas. There is no control valve in the vacuum line,

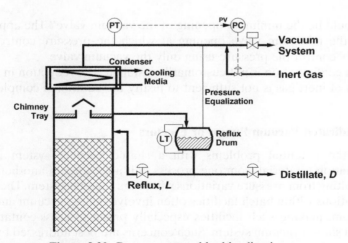

Figure 3.20. Pressure control by bleeding inerts.

only a block valve. The major advantage of this configuration is that it will provide good vacuum control even when the noncondensables flow from the tower is zero.

In some processes, the presence of oxygen leads to many consequences, most being rather unpleasant. Of course, in-leakage of air into a process under vacuum is one source of oxygen. Instead of admitting the inert gas immediately upstream of the vacuum control valve as in Figure 3.20, the inert gas should be admitted into the process. For vacuum towers, bleeding the inert gas into the reboiler would sweep some of the oxygen out of the tower and into the vacuum system.

A common concern with the configuration in Figure 3.20 is the rate of consumption of inert gas. The cost depends on the nature of the inert gas, but even for an inert such as nitrogen, the rate of consumption is usually acceptable.

The consumption of inert gas can be reduced by implementing a split range configuration utilizing two control valves:

Control valve on inert gas bleed. This valve is used to control the tower pressure when the flow of noncondensables is near zero. This valve operates when the pressure controller output is below mid-range. As the pressure controller output changes from 50% to 0%, the inert gas valve opening changes from 0% to 100%.

Control valve on the vacuum line. This valve can be used to control the tower pressure except when the flow of noncondensables is near zero. When the pressure controller output is below mid-range, the vacuum valve is at some minimum opening. As the pressure controller output increases from 50% to 100%, the vacuum valve opening increases from the minimum opening to 100%.

What should be the minimum opening of the vacuum valve? The appropriate value is the minimum valve opening at which the pressure controller can effectively control the pressure using only the vacuum valve.

Such a configuration is not customarily installed. The reduction in the consumption of inert gas is not sufficient to justify the additional complexity.

3.8.4. Dedicated Vacuum Pumps or Ejectors

One of the potential problems with a shared vacuum system is cross-contamination (material from one process unit appearing in another process unit) resulting from pressure variations within the vacuum system. The sequential operations within batch facilities often involve pulling vacuum and breaking vacuum, making such facilities especially prone to cross-contamination through a shared vacuum system. Such concerns are best addressed by installing dedicated vacuum systems for each process unit.

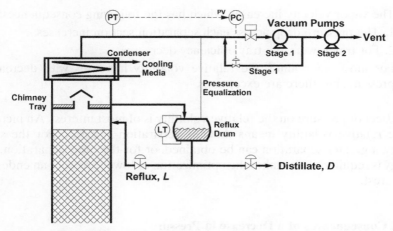

Figure 3.21. Pressure control using recycle around vacuum pump.

When the vacuum is provided by pumps that are dedicated to the tower, the tower pressure can be controlled via the vacuum pump. One approach is to equip the stage 1 vacuum pump with a variable speed drive. However, problems again arise when the flow of noncondensables from the tower is very small.

In the configuration in Figure 3.21, the vacuum is controlled by manipulating a control valve in a recycle line around the stage 1 vacuum pump. In a certain sense, this configuration is equivalent to controlling vacuum using an inert gas bleed, which was illustrated in Figure 3.20. The source of the inert gas is the exhaust of the stage 1 vacuum pump.

A similar approach can be applied to towers for which the vacuum is provided by ejectors that are dedicated to the tower. Usually, there are multiple ejector stages. The vacuum can be controlled by recycling gas from the exhaust of the first ejector stage.

Another potential approach to control the tower pressure is to install a control valve on the steam to the first ejector stage. However, problems arise when the flow of noncondensables is very small.

3.9. FLOATING PRESSURE/PRESSURE MINIMIZATION

Operating a tower under pressure control with the same set point throughout the year is usually not the most efficient approach. A decrease in tower pressure has the following effects (boilup and D/F ratio are constant):

- Temperatures decrease throughout the tower.
- The latent heat of vaporization increases. That means a higher heat duty in both the condenser and reboiler.

- The vapor velocity increases, which has the following consequences:
    1. The pressure drop across each separation section increases.
    2. For tray towers, the tray efficiency decreases.
- For most compounds, the relative volatility increases with decreasing pressure, but there are exceptions.

The effect of pressure on the relative volatility is of most interest. An increase in the relative volatility means that the separation is easier. For the same energy, a greater separation can be obtained, or for the same separation, less energy is required. This makes the optimization of tower pressure an endeavor of interest.

3.9.1. Consequences of a Decrease in Pressure

A decrease in pressure has the following effects on tower operations:

- The ΔT in the condenser decreases. During the summer months, this is not advisable. But what about operating at a lower pressure during the winter and at a higher pressure during the summer?
- The ΔT in the reboiler increases. Sometimes, the temperature of the heat source imposes the upper limit on the column pressure.
- Tower flooding is a possibility, depending on the magnitude of the following effects:
    1. The higher vapor velocity at the lower pressure increases the ΔP across each separation section.
    2. If the relative volatility is higher at the lower pressure, then the boilup and reflux can be reduced.
- When inferring composition from temperature, changes in the column pressure is a complicating factor. However, techniques such as compensating the temperature measurements for pressure can be applied.

The key is the impact of pressure on the relative volatility. If the relative volatility decreases with tower pressure, it is usually advantageous to lower the tower pressure to the limit imposed by the condenser. This is the case for most (but not all) towers and is the only case examined herein.

3.9.2. Depropanizer

The values for the base case solution for the depropanizer were presented in Section 1.9. The tower pressure for the base case is 16.0 barg. To determine the effect of pressure on relative volatility, we only need to change the tower pressure and recompute the solution.

Retaining the boilup and product flows for the base case, the following are the results for tower pressures of 15.5 and 16.5 barg:

Column pressure, barg	15.5	16.0	16.5
Heavy key in distillate, mol%	0.0193	0.0283	0.0428
Light key in bottoms, mol%	0.777	0.780	0.796
Boilup, mol/h	64.9	64.9	64.9

The effects of increasing the tower pressure are as follows:

Heavy key in the distillate. Significant increase.

Lignt key in the bottoms. Small increase.

Clearly, the separation is easier at lower pressures. There are exceptions, so an exercise such as this is necessary before embarking on any effort to operate a tower at a lower pressure.

This analysis can be extended to estimate the benefits from decreasing the tower pressure. In the following steady-state solutions, the boilup has been adjusted so that the heavy key in the distillate is the same at the three pressures:

Column pressure, barg	15.5	16.0	16.5
Heavy key in distillate, mol%	0.0283	0.0283	0.0283
Light key in bottoms, mol%	0.785	0.780	0.785
Boilup, mol/h	63.5	64.9	66.2
Change in boilup, %	−1.5%	−	+1.5%

The boilup flow is basically a measure of the energy input to the tower. Reducing the tower pressure by 0.5 barg reduces the energy consumption by 1.5%. The change is relatively small, but the rate of return is huge—the cost to realize this reduction is zero.

As will be noted in the subsequent chapter on tower optimization, other potential benefits eclipse the benefits from energy savings. Suppose the feed rate to a tower is at its maximum because the tower is operating at the maximum vapor rate that the reboiler can deliver. Reducing the pressure permits the tower to operate at a 1.5% lower boilup. However, a far more economically attractive option is to increase the feed rate (by approximately 1.5%) and continue to operate the tower at the maximum possible boilup rate. And in this example, there is another benefit. Lowering the tower pressure decreases the temperature on the process side of the reboiler. If the reboiler is in the heat transfer limited mode of operation, lowering the process temperature increases the maximum boilup that the reboiler can deliver.

3.9.3. Floating Pressure

If the desire is to operate the tower at the lowest possible pressure at all times, there is a simple approach to accomplish this:

- Switch the column pressure controller to manual.
- Open the control valve on the cooling media to 100%. If other mechanisms are used to manipulate the heat transfer in the condenser, set the output to the control valve to whatever value is required to give maximum cooling. This is not always fully open. For example, if the heat transfer rate is manipulated through a hot gas bypass valve, then fully close this valve.

This is not a common approach to operating towers, so objections will arise. But why is it necessary to control tower pressure? The most common concerns are the following:

1. Varying tower pressure complicates the use of temperature to infer composition. However, techniques such as pressure-compensated temperatures address this issue. Another option is to install analyzers; composition analyses are not affected by tower pressure.
2. A change in pressure on either side of a control valve will affect the flow through the valve. If the pressure is permitted to float, then flow measurements and flow controllers should be considered.

Unfortunately, discontinuing pressure control is likely to be controversial. This leads to an alternative known as pressure minimization, which achieves results close to those of floating pressure while retaining tower pressure control.

3.9.4. Valve Position Control

Pressure minimization relies on a methodology known as valve position control. Valve position controllers are an important tool in optimization endeavors. In most applications, they are used to operate a control valve either nearly fully open or nearly fully closed. That is, they cause the process to operate very close to the constraint imposed by the control valve.

Understanding pressure minimization requires an understanding of valve position control. An accompanying book [1] explains valve position control and provides additional examples of its application.

3.9.5. Pressure Minimization

One criticism of floating pressure is that disturbances such as rain showers result in short-term variations in the tower pressure. Pressure minimization has two objectives:

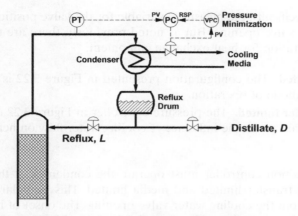

Figure 3.22. Pressure minimization—valve on cooling media.

1. Retain the column pressure controller to respond to the short-term events.
2. Take advantage of long-term effects (such as seasonal temperatures) by adjusting the set point to the pressure controller.

The adjustment of the set point to the pressure controller is provided by the valve position controller illustrated in Figure 3.22. The measured variable for the valve position controller is the cooling media control valve opening (or position signal). The output of the valve position controller is the pressure controller set point.

The valve position controller is tuned to respond very slowly. For short-term disturbances in the pressure, the valve position controller does not significantly change the pressure set point. However, if the pressure controller output remains consistently below the set point for the valve position controller, extra cooling capability is available. The valve position controller slowly lowers the set point to the pressure controller, the objective being to take advantage of the extra cooling capabilities. If the pressure controller output is consistently above the set point to the valve position controller, the valve position controller will slowly increase the pressure set point.

The configuration for the valve position controller depends on the equipment used for the condenser. For the hot gas bypass configurations in Figures 3.10 and 3.11, the input to the valve position controller is the opening of the hot gas bypass valve. The target for the valve position controller is the minimum opening (such as 10%) permitted for this valve.

3.9.6. Heat Transfer Limited

Oversizing is an issue in any valve position control application. The configuration illustrated in Figure 3.22 is the usual formulation for a valve position controller; specifically, the measured variable for the valve position controller is the control valve opening. But as noted previously, there are two possible modes of operation for heat exchange equipment:

Media limited. The configuration presented in Figure 3.22 is appropriate for this mode of operation.

Heat transfer limited. The pressure controller in Figure 3.22 cannot function in this mode; the influence of cooling water flow on heat transfer is too low.

The valve position controller must operate the condenser at the boundary between heat transfer limited and media limited. This boundary cannot be determined from the cooling water valve opening. The onset of heat transfer limited conditions is indicated by the following parameter:

$$\frac{T_{\mathrm{CWR}} - T_{\mathrm{CWS}}}{T_{\mathrm{C}} - T_{\mathrm{CWS}}} = \frac{\text{Temperature rise of the cooling water}}{\text{Maximum } \Delta T \text{ for heat transfer}}.$$

In practice, the overhead vapor temperature T_{OV} can be used in lieu of the condenser temperature T_{C}.

In the configuration in Figure 3.23, this parameter is evaluated and used as the measured variable for the valve position controller. One could question if the name "valve position controller" is still appropriate. The argument for retaining the name is that it more appropriately reflects the function of the controller. The set point for the valve position controller is now the target for the calculated parameter. A value of 0.2 was previously suggested as being the boundary between media limited and heat transfer limited, but a higher value could be used.

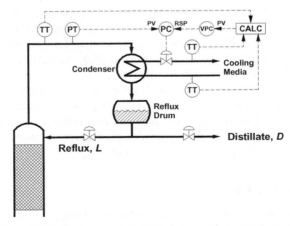

Figure 3.23. Pressure minimization based on temperatures.

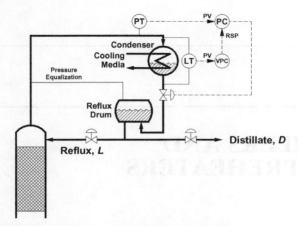

Figure 3.24. Pressure minimization for flooded condenser.

3.9.7. Flooded Condensers

In a flooded condenser, the utilization of the heat transfer capability is most accurately indicated by the level within the condenser. Figure 3.24 presents the pressure minimization configuration when the pressure is controlled by manipulating a control valve on the condensate line between the condenser and the reflux drum (refer to Fig. 3.8).

The percent utilization of the heat transfer capability is related to the condensate level within the condenser. Consequently, the measured value for the valve position controller is the condensate level within the condenser. One could argue that the appropriate name for the controller is a "level controller," but "valve position controller" is used in Figure 3.24 because it more accurately reflects the purpose for the controller.

Figure 3.24 applies specifically to a flooded condenser, but the principles can be generalized. Regardless of the equipment used for the condenser, the measured value for the valve position controller must reflect the percent utilization of the heat transfer capability of the condenser. For various reasons, valve openings may not satisfactorily reflect utilization of the heat transfer capability. Where additional measurements provide a better indication of the utilization of the heat transfer capability, they should be installed.

REFERENCE

1 Smith, C. L., *Advanced Process Control: Beyond Single Loop Control*, John Wiley & Sons, 2010.

4

REBOILERS AND FEED PREHEATERS

Adding heat to a distillation column usually involves some combination of the following:

- Reboiler.
- Feed preheater.
- Economizer.
- Side heater.

The first three are considered in this chapter; the side heater is examined in a later chapter.

Reboilers come in a variety of styles, but they all transfer heat from the heating media to generate vapor within the column. The nature of the heating media always has an impact on the controls:

- **Condensing vapor.** In most cases, the condensing vapor is steam. The discharge is condensate, which must be returned to the steam plant.
- **Hot liquid.** Hot oil systems are quite common in the chemical industry.
- **Fired heater.** These are common in gas processing, refining, and so on.

Feed preheaters are usually installed for one or both of the following reasons:

Maintain constant feed enthalpy. The feed is usually partially vaporized within the feed preheater, which raises issues when temperature is used as the measure of enthalpy.

Distillation Control: An Engineering Perspective, First Edition. Cecil L. Smith.
© 2012 John Wiley & Sons, Inc. Published 2012 by John Wiley & Sons, Inc.

Reduce heat transfer requirements in the reboiler. Being at a lower temperature, heat is more easily transferred to the feed than to the bottoms via the reboiler.

An economizer is basically a feed preheater whose heating media is the bottoms stream.

4.1. TYPES OF REBOILERS

From a control perspective, there are two issues that pertain to the type of reboiler:

Reboiler dynamic response. How quickly does a change in the heat input translate to a change in the boilup? If the tower is operated at constant heat input, this aspect is of no significance. But if either the bottoms level or the bottoms composition is controlled by manipulating the boilup, a rapid response is desirable.

Tower dynamic response. The various liquid holdups within the tower affect the dynamic response of product compositions. Especially in packed towers, the holdups associated with the reboiler and condenser are significant in comparison to the holdup within the tower. Large holdups mean a slow response of the product composition to changes in the manipulated variables (including boilup).

Towers that respond slowly give some operations personnel the impression that the tower is more stable. However, this is not necessarily the case—the result could be a persistent cycle with a period of days. In addition, towers that respond slowly are equally slow to recover from major upsets. The holdup should be no more than what is required by the geometry of the reboiler and related considerations. Steps can be taken to make the bottoms holdup as small as possible, especially when thermal degradation of the material within the tower is a concern.

4.1.1. Kettle Reboiler

As illustrated in Figure 4.1, a weir inside the kettle reboiler assures that the horizontal tube bundle is submerged at all times. The liquid from the tower flows into the reboiler upstream of the weir, some of it is vaporized to provide the boilup, and the remainder flows over the weir to provide the bottoms stream. In a kettle reboiler, the vapor–liquid disengagement is provided within the reboiler. For all other reboiler types, vapor–liquid disengagement occurs within the column.

The bottoms level measurement is also associated with the kettle reboiler. The level transmitter senses the level on the downstream side of the weir. The

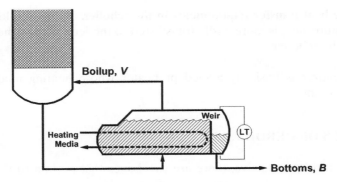

Figure 4.1. Kettle reboiler.

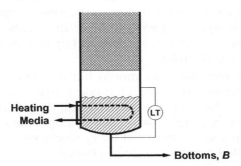

Figure 4.2. Internal exchanger.

liquid volume upstream of the weir is fixed; the variable liquid volume is downstream of the weir.

The kettle reboiler retains a relatively large quantity of liquid. Its effect on column dynamics is as follows:

- The liquid is essentially at the vapor–liquid equilibrium temperature. Consequently, a change in the heat input is quickly translated to a change in boilup.
- Only a small portion of the liquid volume is variable, so the bottoms level usually responds quickly.
- The composition dynamics are affected by the total volume of liquid within the reboiler.

4.1.2. Internal Exchanger

The heat is supplied through an internal tube bundle inserted into the tower below the lower separation section, as illustrated in Figure 4.2. To minimize the impact on tower height, the tube bundle is normally horizontal and must remain submerged by the reservoir of liquid within the tower. The bottoms level is the height of the liquid within the tower.

In towers used for batch distillation, the tower (usually of small diameter) is mounted on top of a vessel commonly referred to as a "pot." The heat source may be a horizontal or vertical tube bundle, the pot could be jacketed, or both. The volume of the pot determines the size of the batch. However, batch distillations are rarely single-run batches. Starting with an empty pot, the initial batch is charged. Some fraction (typically 75% or so) is boiled off. Then the pot is recharged to the initial level, and again some fraction is boiled off. This sequence is repeated some number of times. Only then is the residue removed from the pot and a fresh batch started.

4.1.3. Gravity Flow-Through External Exchanger

To use the reboiler arrangement illustrated in Figure 4.3, suitable tower internals are required to remove all liquid that flows from the lower separation section. Figure 4.3 illustrates the arrangement for trays; appropriate internals are available for packing as well.

The liquid makes only one pass through the exchanger. The stream exiting from the exchanger is partially vaporized when it enters the tower. One must be careful when using this arrangement for towers with a high boilup ratio (V/B). Since the liquid only makes one pass through the exchanger, a large fraction of the liquid flowing through the exchanger must be vaporized in order to achieve a high boilup ratio.

The tower provides for vapor–liquid disengagement. The vapor phase from the exchanger flows to the lower separation section as the boilup. The liquid phase falls into the liquid reservoir in the bottom of the tower. The

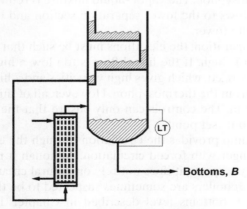

Figure 4.3. Flow-through exchanger.

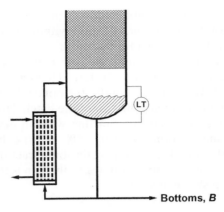

Figure 4.4. Thermosyphon.

amount of liquid within this reservoir determines the measured value for the bottoms level.

With the exchanger external to the tower, steps can be taken to minimize the liquid holdup at the bottoms of the tower. For example, the tower diameter can be reduced (or "necked down") below the vapor return line from the external exchanger.

4.1.4. Thermosyphon

A true thermosyphon reboiler relies entirely on natural recirculation. As illustrated in Figure 4.4, the liquid reservoir is within the tower. The pipe from the bottom of the tower to the inlet to the thermosyphon is entirely filled with liquid. Within the thermosyphon, part, but not all, of the liquid is vaporized, which lowers the liquid head to provide the driving force for liquid to flow through the thermosyphon. The vapor–liquid mixture is returned to the tower, where the vapor flows to the lower separation section and the liquid returns to the bottom of the tower.

For successful operation, the elevations must be such that the flow through the thermosyphon is high. If the liquid flow is too low, a high portion of the liquid must be vaporized, which gives high velocities and a high pressure drop in the return line from the thermosyphon. However, all of this depends on the design of the system. The controls can only ensure that the bottoms level is reasonably close to its set point.

Sometimes a pump provides the circulation through the exchanger. This is an external exchanger with forced circulation. Although it resembles a thermosyphon in many respects, it does not rely on natural circulation.

Thermosyphon reboilers are sometimes suspected to be the culprit for the inverse response in bottoms level described in Chapter 1. Thermosyphon reboilers have some similarities to steam boilers, many of which exhibit inverse response in boiler drum level. But the inverse response in steam boilers is triggered by a sudden drop in pressure associated with an increase in steam demand (somebody opens a very large steam valve). The water within the boiler tubes is either at or near its boiling point. The drop in drum pressure causes some of this water to flash, which displaces water from the boiler tubes into the drum and causes the drum level to increase.

In towers, pressure is not the instigator. When inverse response in bottoms level is observed in towers, the pressure on the process side is usually constant. The explanation based on the thermosyphon as the culprit is as follows:

- Increasing the heat input increases the percent vaporization of the liquid flowing through the thermosyphon tubes.
- The increased vaporization displaces some liquid from the thermosyphon to the bottom of the tower, thus increasing the bottoms level.

If the return line from the thermosyphon is undersized, this effect would be enhanced.

The arguments against this explanation include the following:

- The inverse response occurs too slowly. A change in the heat input should quickly affect the vaporization rate, and hence the appearance of the inverse response.
- For columns that exhibit inverse response to the degree reported by Buckley et al. [1], the amount of liquid in the thermosyphon is too small.
- Inverse response has been observed in columns with other types of reboilers, suggesting that the reboiler is not always the culprit.
- Inverse response has not been reported for packed towers, which reinforces the suspicion that the trays are somehow involved.

4.2. STEAM-HEATED REBOILERS

The most common heating medium for reboilers is steam. Steam supplies are categorized approximately as follows:

Low pressure. Up to 3.5 barg (50 psig); temperatures up to 150°C.

Medium pressure. Above 3.5 barg but less than 17.5 barg (250 psig); temperatures up to 200°C.

High pressure. Above 17.5 barg, but usually not exceeding 40 barg (600 psig); temperatures up to 250°C.

Most plants have at least medium pressure steam; large plants generally have high pressure steam (although not necessarily 40-barg steam).

The temperature on the process side of the reboiler determines the pressure required for the steam supply to the reboiler. Most specialty batch plants have medium pressure steam, which permits the process-side temperature for a reboiler to be up to approximately 180°C (200°C steam less 20°C temperature difference for heat transfer). However, such plants are likely to require a temperature higher than 180°C in one or more reboilers. The alternative is hot oil, which will be discussed in the next section.

As steam is the most common heating medium for reboilers, the discussion in this section is directed specifically at steam. However, the discussion is easily generalized to any condensing vapor.

One attractive feature of heating with steam is the simple relationship between the heat transfer rate and the steam flow. The heat transfer rate is directly proportional to the steam flow, the relationship being approximately

$$Q = F \lambda_s,$$

where

Q = heat transfer rate;

F = steam flow;

λ_s = latent heat of vaporization of the steam.

Sources of errors include the heat losses from the reboiler, variations in condensate temperature, and superheat of the supply steam. These are normally small.

Providing a flow controller for the steam is very common. Maintaining a constant steam flow provides a constant heat input rate, which in a continuous tower usually provides a constant boilup.

4.2.1. Valve on Steam

Although the primary focus must be on delivering the required heat to the reboiler, an important secondary issue is condensate return. Only when the steam consumption is small can the condensate simply flow to a drain. Otherwise, the condensate must be returned to the steam plant.

In the configuration in Figure 4.5, the heat to the reboiler is controlled by manipulating the valve on the steam supply to the reboiler. Changing the steam valve opening changes the steam pressure within the reboiler, which changes the temperature difference for heat transfer.

With the control valve on the steam supply, condensate return issues impose a minimum on the heat transfer rate that can be sustained. Furthermore, this minimum is not a heat transfer rate of zero and a fully closed steam supply valve. As the steam valve opening decreases, the pressure in the shell of the reboiler decreases. What if the pressure in the shell of the reboiler drops below atmospheric? The condensate will not even flow to a drain, but instead is

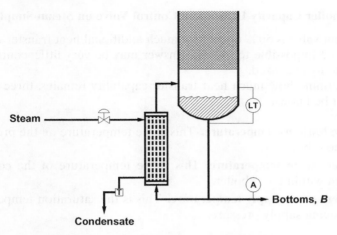

Figure 4.5. Control valve on steam supply.

retained within the reboiler. If the condensate must be returned to the steam plant, this occurs at some pressure above atmospheric.

Assuming the condensate exits at atmospheric pressure, the minimum possible temperature within the reboiler steam chest is 100°C. The driving force for heat transfer is this temperature less the temperature on the process side of the reboiler. The minimum heat transfer rate that can be sustained corresponds to this temperature difference. A controller of some type (usually either steam flow or bottoms level) is manipulating the opening of the control valve. What if the controller determines that a lower heat transfer rate is required? Starting with the steam chest empty of condensate, the entire heat transfer surface area is exposed, which gives a higher heat transfer rate than required. The following cycle ensues:

- The heat transfer rate is too high.
- The controller decreases the steam valve opening.
- When the pressure in the reboiler steam chest drops below the pressure required for condensate return, the reboiler begins to fill with condensate.
- The heat transfer rate decreases.
- The controller increases the steam valve opening.
- Once the pressure in the steam chest exceeds the pressure required for condensate return, the condensate is forced out of the reboiler.
- The heat transfer is again too high, and the cycle repeats.

Often the initial request is to "tune out" this cycle. However, it is a process problem, not a controls problem. Tuning adjustments affect the cycle, but will not eliminate it.

4.2.2. Reboiler Capacity Utilization: Control Valve on Steam Supply

If the steam valve is 50% open, how much additional heat transfer capability is available? Impossible to say. The answer may be very little; control valves are commonly oversized.

To determine how much heat transfer capability remains, three temperatures must be known:

Column bottoms temperature. This is the temperature on the process side of the reboiler.

Reboiler steam temperature. This is the temperature of the condensing steam within the reboiler.

Reboiler steam supply temperature. This is the saturation temperature at the steam supply pressure.

The current ΔT in the reboiler is the difference between the reboiler steam temperature and the bottoms temperature. The maximum possible ΔT in the reboiler is the difference between the steam supply temperature and the bottoms temperature. The ratio of these two ΔT's is the fraction of the reboiler capacity that is currently being used. This makes one assumption: With the control valve fully open, the steam pressure within the reboiler will be the steam supply pressure. This is not guaranteed, but oversizing control valves is so common that this assumption is usually good.

4.2.3. Steam Flow Controller

In Figure 4.6, the steam flow is measured and a flow controller is configured. The measurement range for a flow is usually as follows:

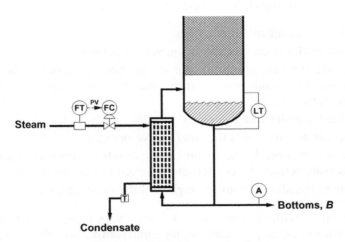

Figure 4.6. Steam flow controller.

Lower range value. Zero.

Upper range value. Somewhat higher than the maximum anticipated flow.

This range is used for the set point as well, so the process operator can specify any value between these two limits.

What if the process operator specifies a value for the steam flow set point that is less than the minimum as explained above? The steam flow cycles as described above. Although the average flow for the cycle will be approximately the steam flow set point, a similar cycle will exist in the boilup, which in turn affects the separation in a similar manner.

What if the process operator specifies a value for the steam flow set point that is greater than the maximum as explained above? The flow controller in Figure 4.6 responds quickly, so the controller rapidly drives the steam valve fully open. This gives a steam flow less than the set point, but there is no cycle.

In some configurations, the steam flow controller is the inner loop of one of the following cascades:

- Bottoms level to steam flow cascade.
- Bottoms composition to steam flow cascade.

What if the outer loop of one of these cascades specifies a steam flow set point higher than can be achieved? There are two consequences:

1. The steam flow controller will drive the steam valve fully open. The customary windup protection will be invoked within the flow loop.
2. The outer loop of the cascade will continue to increase the steam flow set point, which is a form of windup. To prevent this from occurring, the cascade must be appropriately configured so that one of the following windup protection mechanisms is invoked:

   External reset. For conventional controls, this was the only option. Some, but not all, digital control systems provide this capability.

   Integral tracking. When the flow controller output attains the upper output limit, the integral mode in the outer loop must track the measured value of the steam flow. This effectively disables the integral mode in the outer loop.

   Inhibit increase. When the flow controller output attains the upper output limit, the outer loop is not allowed to further increase its output, which is the steam flow set point for the inner loop.

These mechanisms are explained in an accompanying book [2].

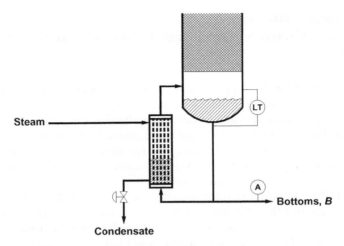

Figure 4.7. Control valve on condensate.

4.2.4. Valve on Condensate

In Figure 4.7, the control valve is on the condensate. Some condensate is retained within the reboiler. The effective area for heat transfer is the heat transfer surface area exposed to the condensing vapor. The submerged heat transfer area merely subcools the condensate, with little contribution to the total heat transfer.

The control valve on the condensate will be smaller than the control valve on the steam, which reduces its cost. However, the main incentive to use the configuration in Figure 4.7 pertains to condensate return. The steam pressure within the reboiler is always the steam supply pressure. Consequently, the full steam supply pressure provides the driving force for condensate return.

For the configuration in Figure 4.7, the minimum heat transfer rate is theoretically zero. That is, the reboiler could completely fill with condensate, which would reduce the heat transfer rate to zero.

However, problems arise should the maximum heat transfer rate be attained. As the control valve opens, the condensate level within the reboiler drops. In most applications, the control valve can be opened sufficiently to force all condensate out of the reboiler. Steam flows through the condensate valve and into the condensate return system. The reboiler is said to "blow steam," which must not be allowed to happen. Not only are there consequences within the condensate return system; the steam pressure within the reboiler drops, which results in a major upset to the tower.

4.2.5. Valve on the Condensate with Trap

Figure 4.8 presents a simple way to prevent steam from being released into the condensate return system: Install a steam trap upstream of the condensate

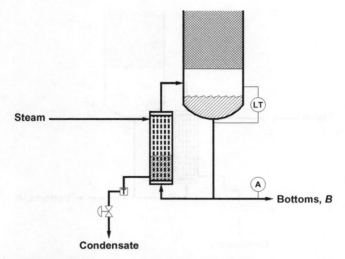

Figure 4.8. Steam trap upstream of control valve on condensate.

control valve. Under normal conditions, the control valve causes condensate to be retained within the reboiler. As the trap passes condensate, it has no effect. But should all condensate drain from the reboiler, the trap will prevent steam from flowing into the condensate return system.

The steam trap prevents the reboiler from "blowing steam," but there is a side effect. Once the trap begins to block steam, further opening the control valve has no effect on anything (steam flow, bottoms level, heat transfer rate, etc.). The controller that outputs to the condensate valve will drive the valve fully open. Only then are the customary windup protection mechanisms invoked. This is too late; windup protection should be invoked when the controller output ceases to have any effect on the measured variable. This occurs when the steam trap begins to block steam, but normally the control system has no indication that the trap is blocking steam.

4.2.6. Flow Control with the Condensate Valve

Figure 4.9 illustrates the use of a flow controller when the valve is on the condensate. The advantages are the same as when the valve is on the steam. The heat input to the reboiler is directly proportional to the steam flow, whereas the relationship between heat input and the condensate control valve position is complex and nonlinear.

A word of caution in tuning the flow controller. Most organizations have recommended values for the tuning coefficients for flow controllers. But there is a caveat to these recommendations that is too frequently omitted. The recommendations apply when the measured flow is the flow through the control valve. This is the case when the control valve is on the steam supply as in Figure

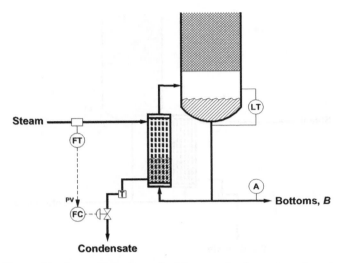

Figure 4.9. Steam flow control with control valve on condensate.

4.6, but this is not the case when the control valve is on the condensate as in Figure 4.9. The measured flow is the steam flow; the control valve is on the condensate flow. The flow controller in Figure 4.9 will respond far more slowly than the typical flow controller. Changing the condensate valve opening does not immediately affect the steam flow. Instead, changing the control valve opening causes the condensate level within the reboiler to change, which affects the heat transfer and the steam flow.

If the condensate drains completely out of the reboiler and the trap blocks the steam, the condensate control valve has no effect on the steam flow and the flow controller winds up. The typical flow loop is fast and will unwind quickly, but the flow loop in Figure 4.9 is far slower than the typical flow loop.

4.2.7. Steam Flow Control with Condensate Level Override

The override configuration illustrated in Figure 4.10 is one approach to avoid windup in the flow controller. The override configuration functions as follows:

- If the condensate level is above its minimum value (as determined by the set point to the condensate level controller), the steam flow controller determines the opening of the condensate valve. As the condensate level is above its set point, the condensate level controller will increase its output as much as permitted.
- If the condensate level is at its minimum value, the condensate level controller determines the opening of the condensate valve. The steam flow will be below its set point, so the steam flow controller will increase its output as much as permitted.

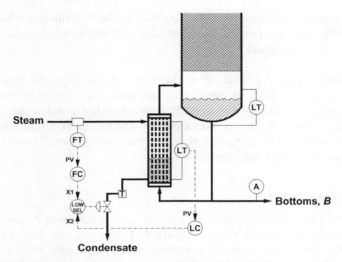

Figure 4.10. Steam flow control with condensate level override.

The low selector switches between steam flow control and condensate level control based on the outputs of the two controllers.

The selector is a bit more sophisticated than suggested by the illustration in Figure 4.10. Unless appropriate logic is incorporated into the configuration, the controller that is not determining the opening of the condensate valve will drive its output to the upper output limit, usually 100% or slightly higher. If the controllers are allowed to do this, the result is windup and the switch between steam flow control and condensate level control will not be smooth. There are three options for preventing this windup:

- External reset.
- Integral tracking.
- Inhibit increase/inhibit decrease.

Override control is explained in an accompanying book [2] along with the mechanisms to prevent windup in override control configurations.

4.2.8. Reboiler Capacity Utilization: Control Valve on the Condensate

If the condensate valve is 50% open, how much additional heat transfer capability is available from the reboiler? The utilization of the heat transfer capability cannot be determined from the condensate valve opening.

Installing a condensate level measurement offers two benefits:

1. The override configuration in Figure 4.10 can be implemented to prevent windup.

2. The capacity utilization for the reboiler can be determined, which makes possible optimization in the form of constraint control (described in a subsequent chapter).

The reboiler capacity utilization is the ratio of the exposed heat transfer surface area to the total heat transfer surface area. For vertical exchangers the reboiler capacity utilization is linearly related to the condensate level. For horizontal exchangers, the reboiler capacity utilization as a function of level is normally provided by a characterization function computed from the physical dimensions of the exchanger.

4.2.9. Condensate Pot

Condensate pot configurations also prevent a reboiler from releasing steam into the condensate return system. Figure 4.11 presents one design for a condensate pot; there are numerous variations. The pressure within both the reboiler and the condensate pot is the steam supply pressure, which means that the full steam supply pressure provides the driving force for condensate return. Condensate drains from the reboiler through the control valve and into the condensate pot by gravity; the designers must ensure the proper elevations for both the reboiler and the condensate pot. The condensate level in the condensate pot is measured and controlled by manipulating the control valve on the condensate return.

The control valve in the condensate line from the reboiler to the condensate pot determines the heat transfer in the reboiler. Changing the opening of this valve affects the condensate level in the reboiler, which affects the heat

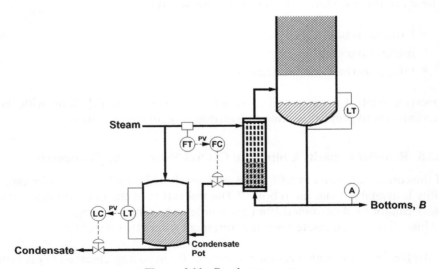

Figure 4.11. Condensate pot.

transfer. Although not required, the configuration in Figure 4.11 provides for flow control of the steam to the reboiler.

In Figure 4.11, only one reboiler discharges into the condensate pot. To make the installation of the condensate pot more cost-effective, several reboilers and possibly other steam-heated exchangers can discharge into a common condensate pot. But to share the same condensate pot, the physical elevations of the various reboilers and exchangers must be correct.

4.3. HOT OIL

Hot oil systems are used when the required temperature of the heat source to the reboiler is above what can be attained with the highest pressure steam available at the plant. Hot oil supply temperatures up to 300°C are common; temperatures up to 400°C are possible.

Several manufacturers produce packaged hot oil systems that include a hot oil heater and a reservoir of hot oil that can be supplied to several exchangers. The hot oil heater attempts to maintain a constant temperature for the hot oil supply. However, at times of peak demand, the heating capability of the hot oil heater can be exceeded, and the hot oil supply temperature will drop.

4.3.1. Once-Through Arrangement

Figure 4.12 presents the once-through configuration. The hot oil makes a single pass through the reboiler and exits to the hot oil return. In the previous chapter devoted to condensers, the characteristics of once-through heat transfer arrangements were examined. The relationship of heat transfer to the hot oil flow is similarly complex. But with hot oil, oversizing the capability to flow hot oil through the exchanger is less common.

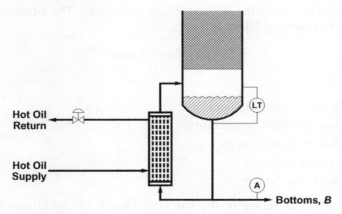

Figure 4.12. Once-through hot oil arrangement.

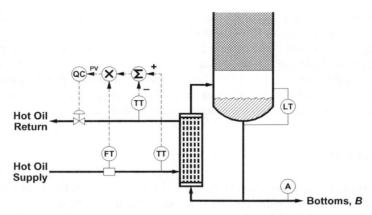

Figure 4.13. Btu controller for once-through hot oil arrangement.

Especially in batch facilities that impose a variable demand on the hot oil supply system, two disturbances often occur during peak demands:

Hot oil supply pressure. Sensing the hot oil flow and configuring a hot oil flow controller will provide very effective response to this disturbance. However, maintaining constant hot oil flow does not assure a constant heat transfer rate.

Hot oil supply temperature. A configuration commonly referred to as a Btu controller should be considered.

4.3.2. Btu Controller

The Btu controller configuration illustrated in Figure 4.13 significantly improves the response to changes in the hot oil supply temperature. The Btu controller computes the heat transfer rate and then manipulates the hot oil control valve to obtain the desired heat transfer rate. The Btu controller is based on the following equation:

$$Q = F\, c_P (T_{SUP} - T_{RTN}),$$

where

T_{RTN} = hot oil return temperature;
T_{SUP} = hot oil supply temperature;
c_P = hot oil heat capacity;
F = hot oil flow;
Q = heat transfer rate.

The obvious disadvantage of the Btu controller is the additional measurements: hot oil flow, hot oil supply temperature, and hot oil return temperature.

Computing the hot oil temperature change entails subtracting two numbers (the hot oil supply temperature and the hot oil return temperature) to obtain a small one (the temperature rise ΔT). Numerically, this is not a good practice. Fortunately, excessive hot oil flow rates and the associated small temperature rises are not as common for hot oil as for cooling water. However, directly sensing the temperature rise is advisable.

Operations personnel often prefer a slightly different but equivalent implementation of the Btu controller:

- Configure a hot oil flow controller.
- Compute the hot oil flow set point F_{SP} to give the desired heat transfer rate Q_{SP}:

$$F_{SP} = \frac{Q_{SP}}{c_P(T_{SUP} - T_{RTN})}.$$

4.3.3. Recirculating Arrangement

In Figure 4.14, the hot oil is being recirculated from the reboiler outlet to the recirculation pump and then to the reboiler inlet. Prior to entering the pump, the hot oil from the supply is combined with the hot oil being recirculated. The hot oil recirculation flow is very high, so the hot oil temperature rise from reboiler inlet to reboiler outlet should be less than 5°C.

In Figure 4.14, a temperature controller maintains the desired temperature in the recirculation loop by manipulating the opening of the hot oil supply valve. The relationship between the heat transfer rate and the hot oil valve opening (and also hot oil flow) is both complex and nonlinear. However, the relationship between the heat transfer rate and the temperature in the recirculation loop is essentially linear.

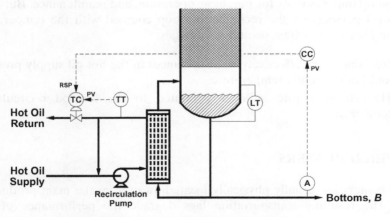

Figure 4.14. Recirculating hot oil arrangement.

To develop a relationship between heat transfer rate and hot oil return temperature, the following assumptions are made:

1. The hot oil in the recirculation loop is at a uniform temperature.
2. The hot oil recirculation temperature is the same as the hot oil return temperature.

With these assumptions, the heat transfer rate is

$$Q = U A (T_{RTN} - T_B),$$

where

T_{RTN} = hot oil return temperature (°C);
T_B = bottoms temperature (°C);
U = heat transfer coefficient (kcal/h m^2 °C);
A = heat transfer area (m^2);
Q = heat transfer rate (kcal/h).

The relationship between the heat transfer rate Q and the hot oil return temperature T_{RTN} is linear.

The temperature control loop in Figure 4.14 can be used as the inner loop in either of the following cascades:

• Bottoms composition to hot oil recirculation temperature.
• Bottoms level to hot oil recirculation temperature.

A level-to-temperature cascade is unusual, but the hot oil recirculation temperature loop in Figure 4.14 responds far more rapidly than the bottoms level.

The main disadvantage of the hot oil recirculation loop is that the recirculation pump imposes costs for purchase, operation, and maintenance. But from a control perspective, the recirculation loop coupled with the temperature loop in Figure 4.14 offers several advantages:

• Responds very effectively to disturbances in the hot oil supply pressure and hot oil supply temperature.
• Heat transfer rate is linearly related to the hot oil recirculation temperature.

4.4. FIRED HEATERS

Fired heaters are usually physically located away from the main production area. The resulting transportation lags degrade the performance of any

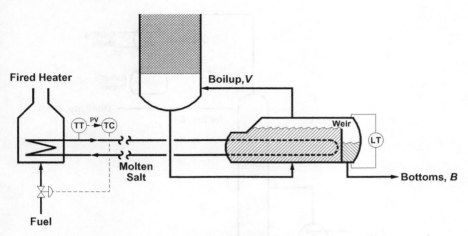

Figure 4.15. Salt bath heater.

controls that manipulate the heat input to the reboiler. The only viable approach is to operate at a fixed heat input rate, making double-end composition control impossible to implement.

Instead of directly heating a flammable organic liquid, an alternative is to use an intermediate fluid that is inflammable. The intermediate fluid is heated in a fired heater to the specified supply temperature. The fluid is then pumped to the process and then returned to be reheated. Even when the fired heater is located away from the main production area, transportation lags are not a problem (transportation lags apply to temperatures, but not to flows). The performance of controls that manipulate the heat input to the reboiler are not degraded, making it possible for double-end composition control.

In some applications, the intermediate fluid is molten salt, and the fired heater is known as a salt bath heater. Figure 4.15 illustrates the arrangement where a salt bath heater supplies the heat to a reboiler. The salt melts at about 200°C, which is the lower limit for the temperature of the heat source. Salt bath heaters are capable of attaining temperatures on the order of 400°C.

The salt bath heater must be physically located a safe distance from the process. The piping distances for the molten salt tend to be long, and flow is by natural convection. These distances coupled with the thermal mass of the molten salt result in a very slow response. The common approach is to control the molten salt outlet temperature by manipulating the firing rate to the furnace. The intent is to supply a constant heat input to the reboiler, which means constant boilup. However, a constant molten salt temperature does not assure a constant heat input at the reboiler.

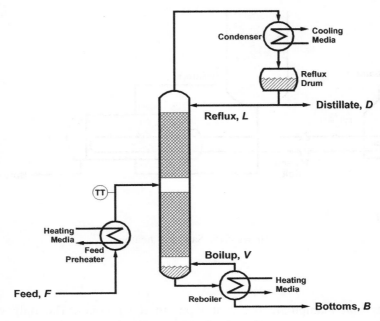

Figure 4.16. Feed preheater.

4.5. FEED PREHEATER

The schematic in Figure 4.16 includes an exchanger known as a feed preheater that adds heat to the feed before it enters the tower. This provides two options for inputting energy to the tower:

- Reboiler.
- Feed preheater.

Upstream of the preheater, the feed is usually a subcooled liquid. The outlet stream from the feed preheater is often partially vaporized, but completely vaporizing the feed is highly unusual. Completely vaporizing the feed leaves any nonvolatile material as a deposit within the equipment.

4.5.1. Feed Preheater versus Reboiler

Now that there are two options for inputting energy into the column, what factors enter the decision as to where to input energy? The two primary factors are the following:

Energy efficiency. Distillation uses energy to obtain separation. If one additional unit of energy is input to the reboiler, what is the improvement in

separation? If one additional unit of energy is input to the feed preheater, what is the improvement in separation? Only a stage-by-stage
separation model can provide quantitative answers. But inputting energy
to the reboiler normally has a greater effect on the separation than inputting energy to the feed preheater. However, until the feed is totally
vaporized, the trade-off is only slightly less than 1:1. Once the feed is
totally vaporized, inputting energy to the feed preheater is much less
effective than inputting the energy to the reboiler.

Energy cost. The bottoms temperature is the highest temperature in the
tower. The bubble point of the feed at the tower pressure is normally
significantly lower. Therefore, to input energy to the reboiler requires a
higher temperature heat source than to input energy to the feed preheater. This is usually reflected in the utility costs. For example, if high
pressure steam is required for the reboiler, there is a good chance that
medium pressure steam can be used in the feed preheater. It is especially
attractive if one can input heat to the feed preheater with low pressure
steam (many facilities have an excess of low pressure steam).

4.5.2. Partial Vaporization in the Feed Preheater

In most applications, enough energy is input to the feed preheater to give a
feed that is partially vaporized upon entry to the tower. The desire is for the
feed preheater to maintain a constant enthalpy for the tower feed, although
the specific value of the feed enthalpy is not critical. That is, errors in the feed
enthalpy can be tolerated provided the value is constant.

The schematic in Figure 4.16 includes a temperature measurement on the
preheater discharge stream. This approach needs to be considered in light of
two approaches for operating the feed preheater:

Maintain a back pressure on the feed preheater. The back pressure is sufficient to prevent any vaporization in the feed preheater. There are two
issues:

1. The power input to the feed pump is larger.
2. For the same enthalpy, the temperature of an all-liquid stream would
   be higher than the temperature of a partially vaporized stream. This
   could require a higher temperature heat source for the feed
   preheater.

Vaporize within the feed preheater. The problem is how to measure the
enthalpy of the preheater exit stream. Some issues arise when using a
temperature measurement.

4.5.3. Feed Enthalpy and Temperature

For the depropanizer described in Section 1.9, the feed composition is 0.4%
ethane, 23.0% propane, 37.0% butane, and 39.6% pentane. If the pressure is

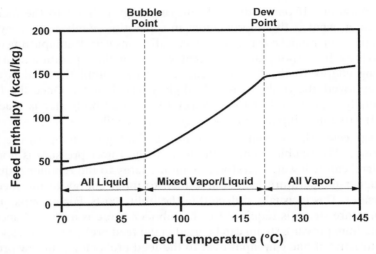

Figure 4.17. Enthalpy of depropanizer feed as a function of temperature.

16.0 barg, Figure 4.17 presents the feed enthalpy as a function of temperature. There are three regions:

Feed temperature less than the bubble point. The feed is entirely liquid. The slope of the graph is the liquid heat capacity. In this region, pressure has no effect on the graph, except that the bubble point increases with pressure.

Feed temperature between the bubble point and dew point. The feed is a mixed vapor and liquid. Most feed preheaters operate in this region. The following two characteristics are significant:

1. For the depropanizer feed, the difference between dew point and bubble point is about 30°C. For feeds with a small difference between dew point and bubble point, the enthalpy increases very rapidly with temperature.

2. Both the bubble point and the dew point depend on pressure. Consequently, the enthalpy in this region is a function of both pressure and temperature. That is, for a given feed enthalpy, the feed temperature increases with pressure. Maintaining a constant feed temperature gives a constant feed enthalpy only if the pressure is constant.

Feed temperature greater than dew point. The feed is entirely vapor. The slope of the graph is the vapor heat capacity. In this region, pressure has no effect on the graph.

4.5.4. Disturbances from the Tower Feed

There are three possible disturbances to the tower that originate with the feed:

Feed flow. This is a very serious upset to the tower. Although there are exceptions, the feed to most towers is controlled by a flow controller. Changes in the flow set point are infrequent and small. Any significant change can be implemented in a ramp fashion.

Feed enthalpy. This upset is almost as serious as a feed flow upset. A change in the feed enthalpy affects both the vapor flow and the liquid flow within the tower. The customary approach is to maintain constant feed temperature with the expectation of maintaining constant feed enthalpy. For a mixed vapor–liquid feed, this is not assured. One should prepare the graph in Figure 4.17 for the feed mixture to determine how sensitive the feed enthalpy is to the feed temperature. The graph can be prepared at different pressures to assess the effect of changes in pressure. One possibility is pressure compensating the preheater temperature measurement in the same manner as the control stage temperatures. However, variations in the feed composition affect the coefficient in the equation used for pressure compensation.

Feed composition. Rarely is a composition analysis available for the feed. Although there may be some opinions, the degree to which feed composition upsets occur for a given tower is usually unknown. But as noted previously, the controls must be capable of effectively responding to such disturbances.

4.5.5. Feed Enthalpy Computer

In most cases, the feed is all liquid upon entry to the feed preheater. When the heat source for the feed preheater is steam, the configuration in Figure 4.18 is possible:

- Measure the temperature upstream of the feed preheater and compute the enthalpy.
- Subtract this value from the desired feed enthalpy to give the required enthalpy increase in the feed preheater.
- Multiply the required enthalpy increase by the feed flow to obtain the required heat transfer in the feed preheater.
- Divide the required heat transfer in the feed preheater by the latent heat of vaporization of the steam to obtain the steam flow.

These computations are provided by the enthalpy computer in the configuration in Figure 4.18.

One obvious obstacle is the number of measurements required by the enthalpy computer. However, there is a less obvious problem. The feed to a

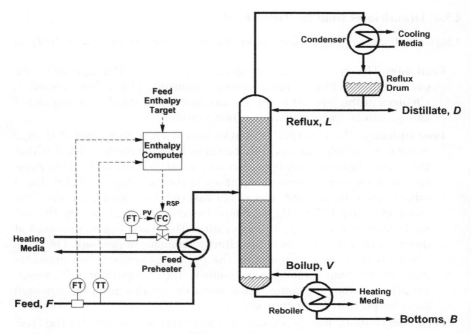

Figure 4.18. Enthalpy computer.

tower is a mixture, and the heat capacity of the feed is a function of the feed composition.

4.6. ECONOMIZER

In many towers, the bottoms product exits at a higher temperature than desired, necessitating an exchanger known as the bottoms cooler to cool the bottoms product to an acceptable temperature. The cooling media may be water or air, but an interesting possibility is to use the tower feed as the cooling media. This is illustrated in Figure 4.19.

Such an exchanger is generally referred to an economizer. Some of the energy that would be lost with the bottoms product is recovered and returned to the tower with the feed. The economizer in Figure 4.19 is equipped with a bypass for the bottoms stream. The maximum amount of heat transfer from the bottoms to the feed occurs when the bypass is closed (or not provided). However, some issues arise when operating with no bypass:

There is no control of the bottoms temperature. If a bottoms color is required to further cool the bottoms stream, the bottoms temperature can be controlled by manipulating the heat removed by the bottoms cooler.

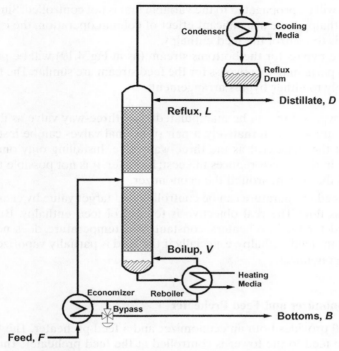

Figure 4.19. Economizer with bypass for bottoms stream.

There is no control of feed enthalpy. Any upset to the bottoms stream (temperature, flow, etc.) affects the enthalpy of the tower feed, which in turn upsets the tower. Basically, an upset can be propagated from the bottoms stream to the feed enthalpy to the internal liquid/vapor flows to the bottoms stream. Consequently, some control of feed enthalpy is normally provided.

4.6.1. Economizer Bypass Arrangements

The two possibilities are the following:

- Bypass part of the bottoms stream, as illustrated in Figure 4.19.
- Bypass part or the feed stream.

In general, there is no inherent advantage of one over the other; however, for a specific application, there could be factors that favor one over the other.

The bypass flow can be manipulated to control either of the following:

- Feed enthalpy.
- Temperature of the bottoms stream at the economizer exit.

All upsets will be propagated to the variable that is not controlled. Since upsets in feed enthalpy have a significant effect of column operation, the customary approach is to control the feed enthalpy.

Only the bypass for the bottoms stream (as in Fig. 4.19) will be presented. The issues pertaining to a bypass for the feed stream are similar. The following points apply to either bypass arrangement:

- The bypass flow can be manipulated via a three-way valve as illustrated in Figure 4.19. Alternatively, a pair of normal valves can be installed for about the same cost as the three-way valve. Installing only one control valve in the bypass reduces the cost; however, it is not possible to bypass all of the stream around the economizer.
- The feed temperature can be controlled to a target value by changing the bypass flow. The real objective is to control feed enthalpy. But as discussed for feed preheaters, constant feed temperature does not assure constant feed enthalpy, especially if the feed is partially vaporized within the economizer.

4.6.2. Economizer and Feed Preheater

Figure 4.20 provides both an economizer and a feed preheater. The temperature of the feed to the tower is controlled at the feed preheater. The options pertaining to the bypass on the economizer are as follows:

With no bypass. The economizer transfers the maximum possible amount of heat from the bottoms stream to the feed stream.

With a bypass. The bottoms temperature at economizer exit can be controlled.

With both an economizer and a feed preheater, the feed will almost certainly be partially vaporized upon entry to the tower. The alternatives were previously discussed for feed preheaters:

- Maintaining sufficient back pressure so that the feed remains all liquid through the feed preheater requires a higher input of power to the feed pump, and increases the temperature on the process side of the feed preheater.
- Allowing the feed to partially vaporize within the feed preheater (and possibly the economizer) raises issues with regard to temperature. Specifically, maintaining constant feed temperature does not assure constant feed enthalpy.

Especially when the feed is partially vaporized, maintaining a constant feed stream temperature at the exit of the economizer does not assure constant

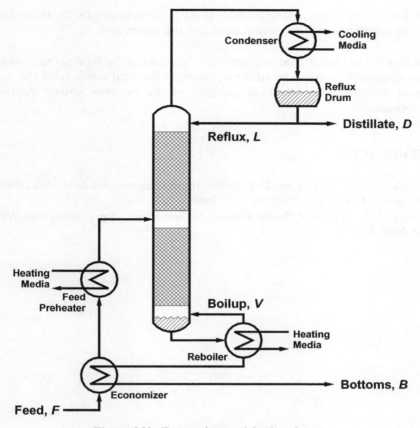

Figure 4.20. Economizer and feed preheater.

feed enthalpy. The feed enthalpy computer presented for a feed preheater in Figure 4.18 can be extended to encompass the economizer. The approach is as follows:

- Measure the feed temperature upstream of the economizer and compute the enthalpy.
- Subtract this value from the desired feed enthalpy to give the required enthalpy increase in the economizer and preheater.
- Multiply the required enthalpy increase by the feed flow to obtain the required heat transfer in the economizer and preheater.
- From measurements of the flow, economizer inlet temperature, and economizer outlet temperature for the bottoms stream, compute the heat transferred to the feed stream in the economizer. Subtract this from the total required heat transfer to give the required heat transfer in the feed preheater.

- Divide the required heat transfer in the feed preheater by the latent heat of vaporization of the steam to obtain the steam flow.

The number of measurements required to implement the feed enthalpy computer is usually an issue. In addition, errors in the heat capacity of the feed stream and errors in the heat capacity of the bottoms stream degrade performance.

REFERENCES

1 Buckley, P. S., R. K. Cox, and D. L. Rollins, Inverse response in a distillation column, *Chemical Engineering Progress*, 71(6), June 1975, 83–84.
2 Smith, C. L., *Advanced Process Control: Beyond Single Loop Control*, John Wiley & Sons, 2010.

5

APPLYING FEEDFORWARD

In process applications, feedforward is often applied to improve the response to disturbances, such as feed flow and feed composition:

Feed flow. The feed flow can usually be measured. Depending on the control configuration, changes in flow can be reflected in the product flows and/or the energy streams (boilup and reflux).

Feed composition. Feed composition analyzers are not routinely installed. But when available, the changes in the feed composition can also be reflected in the product flows.

The third disturbance associated with the feed is enthalpy. The preferred approach is to eliminate such disturbances by maintaining constant feed enthalpy through a feed preheater.

Towers with air-cooled condensers are subject to disturbances related to the weather. A feedforward control technique known as internal reflux control minimizes the effect of a rain event by maintaining a constant internal reflux flow.

What is the most extreme application of feedforward control that has been applied to a distillation column? Despite the fact that it dates from the 1970s,

Distillation Control: An Engineering Perspective, First Edition. Cecil L. Smith.
© 2012 John Wiley & Sons, Inc. Published 2012 by John Wiley & Sons, Inc.

a configuration known as Fractronic [1], if not the winner, is definitely a contender. This configuration has four components:

- Total column material balance.
- Material balance around the vapor space.
- Distillate composition (or upper control stage temperature) control.
- Bottoms composition (or lower control stage temperature) control.

5.1. FEED FLOW AND COMPOSITION

In the configuration in Figure 5.1, the distillate composition controller manipulates the reflux flow. The distillate flow, as determined by the reflux drum level controller, must be the difference between the overhead vapor flow and the reflux flow. The result is an indirect material balance control configuration for distillate composition.

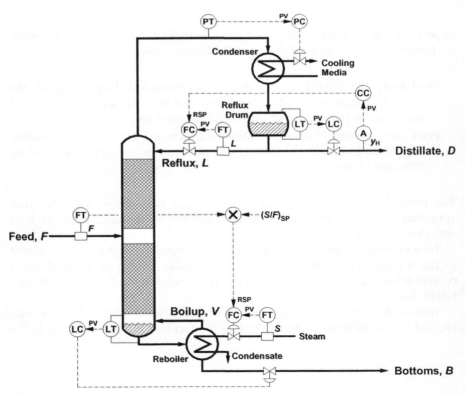

Figure 5.1. Indirect material balance for distillate composition with a steam-to-feed ratio.

Suppose there is a change in the feed flow rate. If a constant boilup is maintained, basically the separation will suffer. The separation is a function of the energy input to the tower, or more correctly, the energy input per unit of feed. On an increase in the feed rate, the following statements apply:

- To maintain the current separation, additional energy is required.
- If the energy input remains constant, the separation will decrease.

All illustrations in this section will be for a water-cooled total condenser with a control valve on the cooling water, only because this is the simplest to draw. However, any of the condenser configurations in Chapter 3 could be substituted.

5.1.1. Steam-to-Feed Ratio

The stage-by-stage separation models usually compute the energy per unit of feed, which suggests that the energy input is directly proportional to the feed flow. This is approximate, but not exact. For example, the heat loss from a column depends primarily on tower temperatures. As the feed rate increases, the issues pertaining to heat losses are as follows:

- The heat loss per unit of feed decreases.
- The appropriate steam-to-feed ratio decreases slightly.

Such impacts are small, but not zero. Hence, feedback trim in some form will be required for any feedforward configuration.

The configuration in Figure 5.1 provides for an automatic ratio of the reboiler steam flow to the tower feed flow. The set point S_{SP} for the steam flow to the reboiler is the feed flow F multiplied by a coefficient a_S:

$$S_{SP} = a_S\, F.$$

Most digital controls also provide a bias coefficient b_S on the ratio:

$$S_{SP} = a_S\, F + b_S.$$

But for the steam-to-feed ratio, the bias b_S should be small and can be set to zero.

The ratio implementation is simplest for a condensing vapor such as steam. For other heat sources, the heat input rate to the reboiler must be ratioed to the feed. Logic is required to calculate the current heat input rate and controls are required to maintain the heat input rate at its target.

5.1.2. Minimum Steam Flow

What if the feed flow F is zero (i.e., column is being operated at total reflux)? The simple ratio in Figure 5.1 would compute a value of zero for the steam flow to the reboiler. This is not acceptable. With no energy input to the reboiler, there is no separation.

For all towers, a minimum must be imposed on the steam flow. The considerations depend on the tower internals:

Tray towers. To prevent excessive weeping of liquid from one tray to the lower tray, the boilup must exceed a minimum that depends on the tray design.

Packed towers. The packing must be wetted at all times, which imposes a minimum reflux flow. The reflux is obtained by condensing vapor, some (and possibly all) of which is produced in the reboiler. A minimum boilup is necessary to provide the required reflux.

Within digital systems, a variety of approaches are available for imposing a minimum on the set point for the steam flow to the reboiler. One possibility is to insert a limiter block between the output of the ratio computation and the remote set point input to the steam flow controller. However, a simpler approach is sometimes possible. For example, some systems include a lower set point limit in the configuration parameter set for the proportional–integral–derivative (PID) block.

5.1.3. Maximum Steam Flow

The heat transfer limitations within the reboiler impose a maximum on the steam flow to the reboiler. What if the ratio calculation generates a value for the steam flow that exceeds this maximum? The flow controller will increase the opening for the steam valve as much as possible, which is usually fully open. The consequences are as follows:

- The actual steam-to-feed ratio is less that the desired steam-to-feed ratio.
- The energy input per unit of feed is lower than desired.
- The separation is lower than desired.
- The impurities in one or both of the product streams increase.

For the control configuration in Figure 5.1, the distillate composition controller will adjust the reflux to maintain the distillate product at its target. However, this is achieved by redirecting more of the light components to the bottoms, the consequences being

- increased amounts of impurities in the bottoms product and
- decrease in recovery (fraction of feed that goes to the distillate product). Light components that should go to the distillate product instead go to the bottoms product.

Two options are available to avoid a loss in the recovery:

- Provide additional heat to the column. This might be possible if the tower is equipped with either a feed preheater or a side heater.
- Reduce the feed flow.

For distillation columns, the process operators are normally expected to take the appropriate action. However, it is possible to automate this using technology similar to the cross-limiters provided for combustion processes. Basically, two ratios are defined:

The desired steam-to-feed ratio. This is used except when the reboiler steam flow is "maxed out."

A minimum steam-to-feed ratio. The feed flow rate is not allowed to exceed the value computed from this ratio (or actually its inverse) and the current steam flow to the reboiler.

Although relatively easy to implement in digital controls, this is not routinely implemented for distillation columns. For combustion processes, the potential consequences, such as a fire in the stack, of inadequate air for the current fuel rate are serious. For distillation columns, the consequences are only economic (loss in recovery).

5.1.4. Reflux-to-Feed Ratio

In the control configuration in Figure 5.1, the distillate composition controller manipulates the reflux flow (indirect material balance). Suppose the feed flow increases by 10%. The ratio controller increases the boilup by approximately 10%. What is the effect on the distillate composition?

The higher boilup means an increase in the overhead vapor flow. There is no immediate change in the reflux flow, which has the following consequences:

- The increase in boilup drives heavy components up the tower. If the reflux is unchanged, the composition of the heavy components increases throughout the tower, including the distillate product.
- The increase in boilup causes the overhead vapor to increase. If the reflux is unchanged, the reflux drum level controller propagates this increase entirely to the distillate flow. This removes more of the light components from the tower, which increases the composition of the heavy components throughout the tower.

An increase in the heavy components causes the impurities in the distillate product to increase. The distillate composition controller responds by

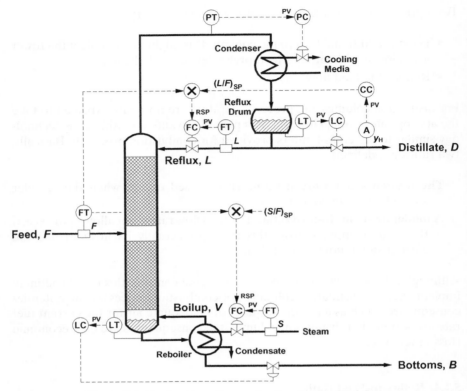

Figure 5.2. Ratio steam flow and reflux flow to feed flow.

increasing the reflux flow, which will return more of the light components to the tower.

When the feed flow changes, only changing the boilup introduces a disturbance to the distillate composition. The distillate composition controller will respond to this disturbance, but this is a slow loop.

Why only ratio the steam to the feed? The configuration in Figure 5.2 also ratios the reflux flow to the feed flow. The set point L_{SP} for the reflux flow is a coefficient a_L times the feed flow F:

$$L_{SP} = a_L\, F$$

A bias for this ratio is normally not required.

The manipulated variable for the distillate composition controller is the coefficient a_L in the ratio equation. Basically, the distillate composition controller is adjusting the desired reflux-to-feed ratio $(L/F)_{SP}$.

A minimum must also be imposed on the reflux flow L. For packed towers, the minimum is the liquid flow required to completely wet the packing. For tray towers, the minimum is usually the reflux flow that corresponds to the minimum boilup (as discussed previously).

For the configuration in Figure 5.2, the permissible range of values for the ratio coefficient a_L is determined by the lower output limit and the upper output limit specified for the distillate composition controller. Large adjustments in this coefficient are not usually required, so a rather narrow range can be specified.

The configuration for the distillate composition controller must also reflect the fact that a minimum is imposed on the reflux flow. Should the reflux flow computed from the ratio computation be less than the minimum reflux flow, adjusting the coefficient a_L has no effect. The distillate composition controller will decrease the value of coefficient a_L to the value specified for the lower output limit in the configuration parameters for the distillate composition controller. Basically, this is reset windup and must be prevented.

The simplest way to prevent this windup is via the inhibit increase/inhibit decrease approach. When the minimum is being imposed on the reflux flow set point, the composition controller should not be allowed to increase its output.

Unfortunately, not all digital controls support inhibit increase/inhibit decrease—the alternatives being integral tracking and external reset. To use these features, the current reflux-to-feed ratio must be computed by dividing the current reflux flow (which is the minimum reflux flow) by the current feed flow. Unfortunately, the feed flow could be zero, which raises the possibility of a division by zero. In practice, the feed flow is likely to be "nearly zero"—a small number that is essentially but not exactly zero. Dividing by such a number produces erratic results. These issues can be addressed, but with some complication to the control logic.

5.1.5. Manipulate Distillate Flow to Control Distillate Composition

In the configuration in Figure 5.3, the distillate composition controller manipulates the distillate flow (direct material balance control). The configuration in Figure 5.3 also includes the steam-to-feed ratio. Some boilup is required when the feed flow is zero, so the issues pertaining to a minimum steam flow discussed above also apply to this configuration.

On an increase in the feed flow, the ratio loop increases the steam flow, which increases the boilup and the overhead vapor flow. But the distillate flow is unchanged (at least in the short term), which has the following consequences:

- The reflux drum level controller returns the increase in the overhead vapor flow to the column as reflux.
- If the feed rate increases but the distillate flow remains the same, more light components are retained within the tower. This increases the

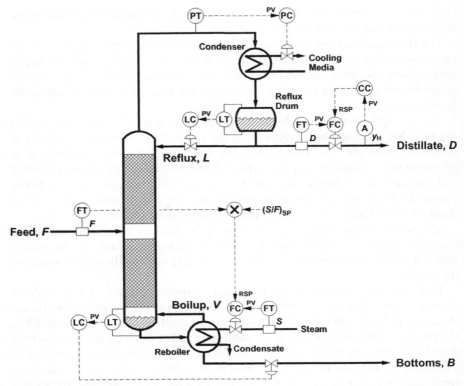

Figure 5.3. Direct material balance for distillate composition with a steam-to-feed ratio.

composition of the light components in all stages, which decreases the heavy components (the impurities) in the distillate product.

The distillation composition controller responds to a decrease in the impurities in the distillate product by increasing the distillate flow. However, this controller responds slowly.

5.1.6. Distillate-to-Feed Ratio

If the feed flow increases by 10%, the distillate flow should also increase by 10% (assuming constant feed composition). This maintains a constant D/F ratio, which means a constant split of the feed into distillate and bottoms products.

The configuration in Figure 5.4 maintains a specified ratio for the distillate flow to the feed flow. The characteristics of this ratio configuration are as follows:

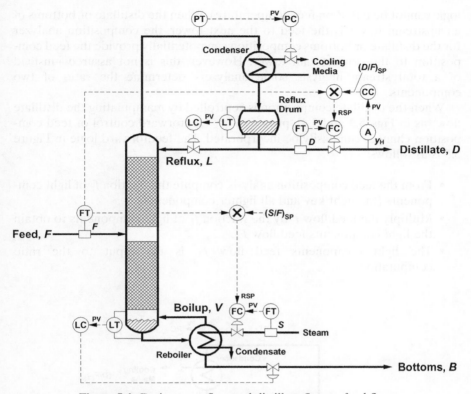

Figure 5.4. Ratio steam flow and distillate flow to feed flow.

- The distillate-to-feed ratio must be a simple ratio:

$$D = a_D \, F.$$

If the feed is stopped, both the distillate flow and the bottoms flow must be stopped.

- With the distillate-to-feed ratio, the manipulated variable for the distillate composition controller is the ratio coefficient a_D. But there is a problem when the feed flow is stopped ($F = 0$). Changing coefficient a_D has no effect on the output of the ratio computation (if $F = 0$, then $D = 0$ regardless of the value of a_D). Either the distillate composition controller must be switched to manual or the output of the controller must be "frozen."

5.1.7. Feed Composition

The main deterrent to feedforward control of feed composition is that composition analyzers are not routinely installed on the column feed (feedforward logic cannot be based on feed temperature). When the distillate or bottoms of an upstream tower is the feed to the next tower, the composition analyzer for the distillate or bottoms composition can potentially provide the feed composition to the downstream tower. However, this is not assured—instead of a total stream analysis, some analyzers determine the ratio of two components.

When the distillate composition is controlled by manipulating the distillate flow (as in Figs. 5.3 and 5.4), provision for feedforward control of feed composition changes can easily be incorporated. The feedforward logic in Figure 5.5 is as follows:

- From the feed composition analysis, compute the fraction f_L of light components (the light key and all lighter components).
- Multiply the feed flow F by the fraction f_L of light components to obtain the light components feed flow F_L.
- The light components feed flow F_L is the input to the ratio computation.

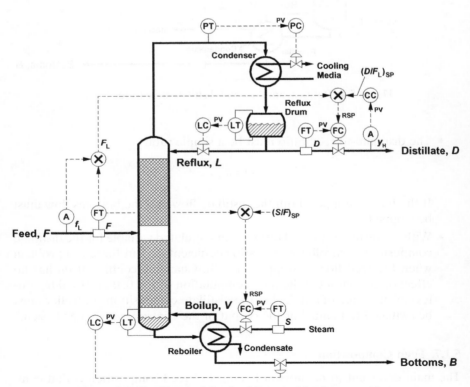

Figure 5.5. Feedforward for feed flow and feed composition.

- The ratio computation is $D = a_D F_L$. In most applications, the coefficient a_D will be close to 1 (occasionally, it is slightly greater than 1).

5.1.8. Dynamic Compensation

If possible, feed flow changes should be implemented on a gradual basis. For example, the operators might make a 5% increase in feed flow by increasing the feed flow by 1% every 6 minutes. Most modern control systems provide a ramp function that facilitates making such changes. Implementing such approaches eliminates the need for dynamic compensation. But in some applications, abrupt changes in the feed rate are unavoidable.

Figure 5.6 is Figure 5.5 with the addition of dynamic compensation elements for the feed flow. Although these elements are indicated as "Lag," they are often implemented using a lead-lag function block with the lead time set to zero. Figure 5.6 includes separate lags for the following:

- The steam-to-feed ratio.
- The distillate-to-feed ratio.

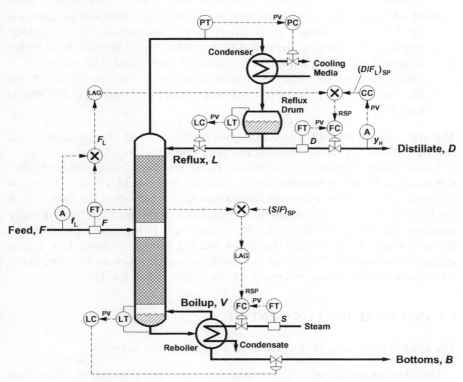

Figure 5.6. Dynamic compensation for feed flow changes.

The dynamics pertaining to each ratio could be different, so this permits different values to be used for the lag times.

Without dynamic compensation, the steam-to-feed ratio configuration will, on a change in the feed flow, immediately change the steam flow. This is usually acceptable in packed towers, but because of the hydraulic lags on liquid flows in tray towers, this response is too fast. The appropriate time to change the steam flow is when the liquid flow increase arrives at the reboiler.

The basis for lagging the flows within a tray tower is to compensate for the hydraulic time constants of the trays. These only pertain to liquid flows; changes in vapor flows are propagated very quickly throughout the tower. The propagation of the liquid flows resulting from a change in the feed rate depends on the nature of the feed:

Feed all liquid. The liquid flow into the lower separation section changes quickly, but the liquid flow out is lagged due to the hydraulic time constants of the trays in the lower separation section. The change in the vapor flow resulting from the change in the heat to the reboiler propagates almost instantly to the top of the tower. The changes arrive at the reboiler and the condenser at almost the same time, so the same lag could be used for both the steam-to-feed ratio and the distillate-to-feed ratio.

Feed all vapor. The vapor flow into the upper separation section changes quickly and is quickly propagated to the condenser. The condenser responds by changing the reflux flow, but the liquid flow to the reboiler is lagged by the hydraulic time constants associated with the trays in both separation sections. No lag is required for the distillate-to-feed ratio, but the lag for the steam-to-feed ratio should be larger than for an all-liquid feed.

The above analysis ignores any dynamics associated with the condenser and reboiler. Mixed vapor–liquid feeds further complicate the analysis.

In practice, the lag time (an adjustable parameter for a lag or lead-lag block) is basically a tuning parameter that is adjusted in the field. For the steam-to-feed ratio, the basis for adjusting this parameter is to observe the response of the bottoms level on an increase in the feed rate. If the bottoms level initially increases, the lag time is too long (the steam needs to increase more rapidly). If the bottoms level initially decreases, the lag time is to short (the steam increases too rapidly). The lag for the distillate-to-feed ratio can be tuned using a similar approach based on changes in the reflux drum level.

5.2. INTERNAL REFLUX CONTROL

The control configuration in Figure 5.7 consists of the following:

Distillate composition. The composition controller manipulates the reflux flow.

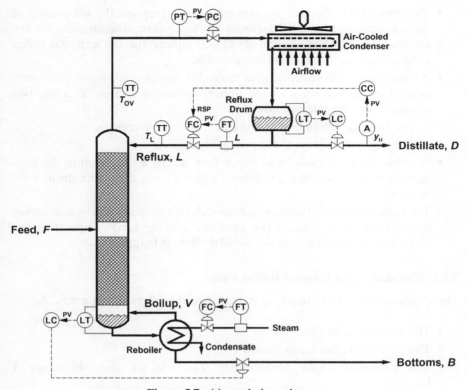

Figure 5.7. Air-cooled condenser.

Reflux drum level. The level controller manipulates the distillate flow.

Column pressure. The heat transfer rate in the condenser is affected through a control valve in the overhead vapor line. On increasing pressure, the pressure controller opens the control valve.

Bottoms level. The level controller manipulates the bottoms flow.

Boilup. The steam flow controller maintains constant boilup.

This configuration is an indirect material balance control configuration for distillate composition.

5.2.1. Rain Event

A major upset to an air-cooled condenser is a rain event. For plants located in arid regions, this upset is even more pronounced. Even a brief rain shower causes the following sequence of events to occur (assuming constant external reflux flow):

- The immediate effect of the rain event is to increase the subcooling of the condensate. Figure 5.7 includes a temperature measurement for the external reflux. Decreases of 25°C per minute for the external reflux temperature on a rain event are possible.
- An increase in the subcooling of the reflux causes more vapor to be condensed in the top stage of the upper separation section. This has two consequences on internal flows:
  1. The internal reflux flow leaving the top stage increases.
  2. The vapor flow from the top stage decreases.
- A reduction in the overhead vapor flow means a reduction in the condensation rate from the condenser, which causes the reflux drum level to drop.
- The reflux drum level controller responds to a drop in level by decreasing the distillate flow. Neither the feed rate nor the feed composition has changed, so this change in the distillate flow is inappropriate.

5.2.2. Calculating the Internal Reflux Flow

Three measurements are required to calculate the internal reflux flow L_I:

- The external reflux flow L.
- The external reflux temperature T_L.
- The overhead vapor temperature T_{OV}. This is also the stage 1 temperature.

When the reflux is subcooled, it must be heated to the temperature of stage 1. The energy Q required to raise the temperature of the external reflux from T_L to T_{OV} is

$$Q = L\, c_P\, (T_{OV} - T_L),$$

where c_P is the heat capacity of the external reflux.

What is the source of the energy to heat the reflux from T_L to T_{OV}? By condensing vapor on stage 1. The amount of vapor condensed L_C is computed as follows:

$$L_C = \frac{L\, c_P\, (T_{OV} - T_L)}{\lambda},$$

where λ is the latent heat of vaporization of the overhead vapor.

The internal reflux L_I is the external reflux L plus the reflux condensed on stage 1:

$$L_I = L + L_C = L + \frac{L\, c_P\, (T_{OV} - T_L)}{\lambda} = L\left[1 + \frac{c_P\, (T_{OV} - T_L)}{\lambda}\right].$$

The internal reflux flow L_1 is the external reflux flow L multiplied by a factor. As the external reflux temperature T_L decreases, the value of the factor increases.

5.2.3. Implementation in Function Blocks

The first step is to rearrange the previous equation to give the ratio L_1/L:

$$\frac{L_1}{L} = 1 + \frac{c_P}{\lambda}(T_{OV} - T_L).$$

Most digital systems permit the internal reflux calculation to be implemented in two function blocks.

Summer. This function block computes the ratio of the internal reflux flow to the external reflux flow. This ratio depends on the overhead temperature and the external reflux temperature, the equation being

$$Y = k_0 + k_1 X_1 + k_2 X_2,$$

where

$$k_0 = 1,$$
$$k_1 = c_P / \lambda,$$
$$k_2 = -c_P / \lambda,$$
$$X_1 = T_{OV},$$
$$X_2 = T_L,$$
$$Y = L_1/L.$$

Multiplier. Multiplying the ratio of internal to external reflux flows by the external reflux flow gives the internal reflux flow. The equation is as follows:

$$Y = X_1 \times X_2,$$

where

$$X_1 = L_1/L \text{ (output of the summer),}$$
$$X_2 = L,$$
$$Y = L_1.$$

If the external reflux flow L is maintained constant during a rain event, the value of the internal reflux flow L_1 increases. But instead of maintaining the external reflux flow L constant, the internal reflux flow L_1 should be constant. This is achieved by reducing the external reflux flow L in a manner that reflects the decrease in the external reflux temperature T_L. There are two control configurations for doing this.

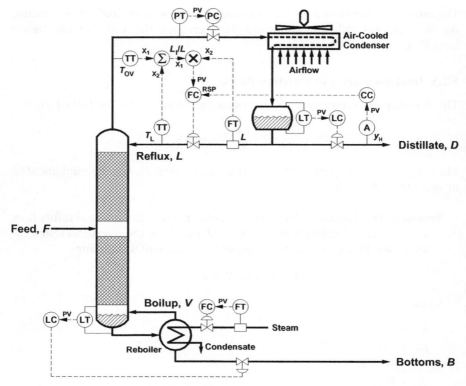

Figure 5.8. Internal reflux flow controller.

5.2.4. Internal Reflux Flow Controller

In the configuration illustrated in Figure 5.8, the computed value for the internal reflux flow L_I is the measured variable for a flow controller. Therefore, the output of the distillate composition controller is the set point for the internal reflux flow. The internal reflux flow controller manipulates the reflux control valve, and consequently the external reflux flow.

Upon a rain event, the internal reflux flow controller responds to the disturbance. The increase in the subcooling (the decrease in the external reflux temperature T_L) causes the computed value of the internal reflux flow L_I to increase above the internal reflux flow set point from the distillate composition controller. However, flow controllers respond very quickly. As soon as the computed value for the internal reflux flow exceeds its set point, the internal reflux flow controller decreases its output, which decreases the reflux valve opening and the external reflux flow L.

Because a flow controller responds so quickly, the internal reflux flow L_I will be maintained very close to its set point throughout the rain event. In

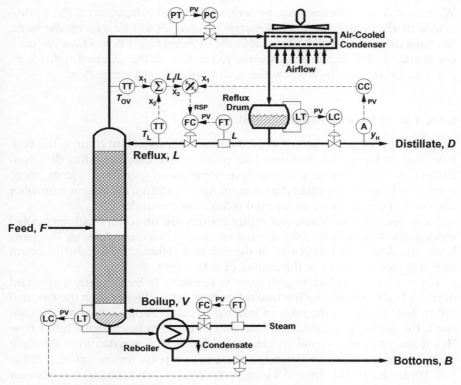

Figure 5.9. Computed set point for external reflux flow controller.

effect, changes in the external reflux temperature T_L are quickly translated into changes in the external reflux flow L.

5.2.5. Computed External Reflux Flow Set Point

The configuration in Figure 5.9 implements internal reflux control as follows:

- The ratio of internal to external reflux flow L_I/L is computed from the external reflux temperature T_L and the overhead vapor temperature T_{OV}.
- The output of the distillate composition controller is considered to be the target for the internal reflux flow.
- The target for the internal reflux flow is divided by the ratio of internal to external reflux flow L_I/L to obtain the target for the external reflux flow.
- The target for the external reflux flow is the set point for the external reflux flow controller.

With regard to variables such as composition and temperature, the performance of the configurations in Figures 5.8 and 5.9 will be exactly the same. Nominal differences will be present in the external reflux flow. However, each configuration will reduce the external reflux flow as the external reflux temperature drops, thereby maintaining a constant internal reflux flow.

5.2.6. Use of Internal Reflux Control

Of the two configurations for implementing internal reflux control, the configuration in Figure 5.9 (computed set point for the external reflux flow controller) is most commonly installed. Operations personnel seem more comfortable with computing the set point for an external reflux flow controller than with the concept of an internal reflux flow controller.

Most installations of internal reflux control are on towers with air-cooled condensers. Rarely is it really needed on water-cooled condensers. The issue is not the degree of subcooling of the external reflux; the issue is the extent to which changes occur in the degree of subcooling.

If the external reflux temperature is constant (which means a constant degree of subcooling), the internal reflux flow will be higher than the external reflux flow. However, the ratio of internal to external reflux flow will be constant. The distillate composition is actually affected by the internal reflux flow. But if the ratio of internal to external reflux flow is constant, the distillate composition controller will merely reduce the set point for the external reflux flow controller by this ratio. As long as the ratio is constant (or changes very slowly), the distillate composition controller functions as expected. Problems arise when the ratio changes rapidly. A rain event with air-cooled condenser does just that.

5.3. EXTREME FEEDFORWARD

As noted in the introduction to this chapter, a column control configuration known as Fractronic [1] takes feedforward control to the extreme. One of the stated objectives for Fractronic was to advance the control of distillation columns as far as possible without resorting to composition analyzers. However, the same approaches can be taken with analyzers; so to be consistent, the discussion herein will be based on composition control rather than stage temperature control.

Fractronic was developed within the American Oil organization over 30 years ago and predates the widespread installation of digital controls. The original Fractronic systems were constructed using rack-mounted electronic analog modules that were preassembled and delivered to the plant as a packaged system. The advantages of using digital technology to implement such a configuration should be obvious.

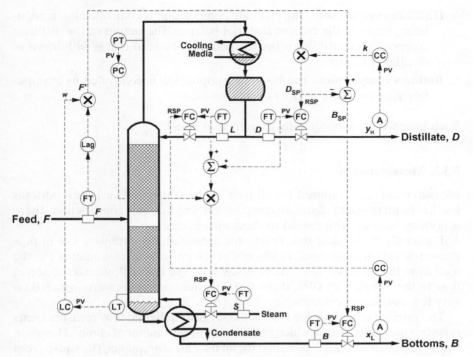

Figure 5.10. Fractronic configuration.

Figure 5.10 presents the Fractronic configuration but using composition control instead of stage temperature control. Both distillate composition and bottoms composition are being controlled, resulting in double-end composition control.

5.3.1. Control Loops

The tower in Figure 5.10 is equipped with a skintight reflux drum and a flooded condenser. Consequently, there are four variables to control:

Bottoms level. Fractronic provided two different mechanisms for bottoms level control, one for when the bottoms level is close to the target and one for when the control error in the bottoms level is large. The one illustrated in Figure 5.10 is used when the bottoms level is close to the target.

Column pressure. The column pressure contains some feedforward components, but ultimately the column pressure is controlled by manipulating the steam flow to the reboiler.

Distillate composition. The distillate composition is controlled by manipulating terms in the column material balance. The output of the distillate composition controller is the fraction k of the feed that is withdrawn as distillate product.

Bottoms composition. The bottoms composition is controlled by manipulating terms in the column energy balance.

Each loop will be subsequently examined in detail.

5.3.2. Measurements

Measurements are required for all four of the controlled variables: bottoms level, column pressure, distillate composition, and bottoms composition. There is nothing unusual with regard to these measurements.

Especially at the time that Fractronic appeared, the extensive use of flow measurements was unusual. Fractronic relies on flow measurements for the feed flow, the distillate flow, the bottoms flow, the reflux flow, and the steam flow to the reboiler. In 1976, these five flow measurements were installed on very few distillation columns.

The original schematic for Fractronic suggested that all flow measurements relied on the orifice meter, which in 1976 was the flow meter of choice. However, square root extractors are provided for all flow measurements. The square root extractor is not required for flow control; flow controllers respond so rapidly that linearization of the flow measurement has little impact on flow controller performance. But for the feedforward functions, the linearization provided by the square root extractor is essential. When summing flows, computing set points for flows, and so on, the flow measurement must be linearized.

5.3.3. Complexity

Certainly the first impression of the control configuration in Figure 5.10 is extreme complexity. This is partially due to the nature of piping and instrumentation (P&I) diagrams. They work very well for simple loop configurations. But as components are added for functions such as feedforward control, the complexity of a P&I diagram increases rapidly.

This is unfortunate. The configuration in Figure 5.10 is really not very complex, which will become more evident when each loop is examined individually. In addition to the components of the four flow loops, the P&I diagram for Fractronic contains only the following simple components:

- Four loop controllers (bottoms level controller, column pressure controller, distillate composition controller, and bottoms composition controller). These controllers would be required in a simple feedback configuration.

- Four summers.
- Four multipliers.

Certainly not a challenge for a digital control system.

5.4. FEEDFORWARD FOR BOTTOMS LEVEL

The schematic in Figure 5.11 contains only those components of Fractronic that pertain to controlling the bottoms level. Briefly, the approach is as follows, starting at the feed flow measurement:

- A lag is applied to the feed flow measurement to provide dynamic compensation. The need for this was previously discussed in connection with ratioing steam to feed and a product flow to the feed.
- The multiplier applies a correction factor w to the feed flow measurement to obtain the corrected feed flow measurement. The following notation is used:

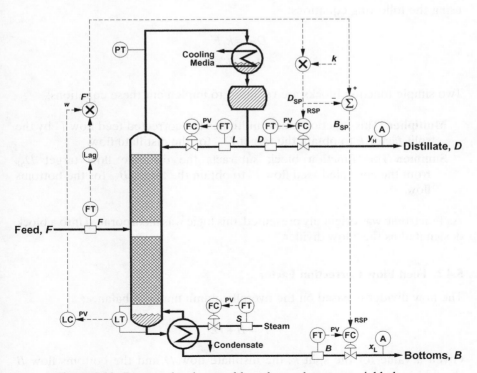

Figure 5.11. Bottoms level control based on column material balance.

F = measured value for the feed flow;

w = feed flow correction factor;

F' = $w\, F$ = corrected value for the feed flow.

How this correction factor is obtained will be discussed shortly.

- Targets for the distillate flow and the bottoms flow are computed from the corrected feed flow F' and the fraction k for the split:

  k = fraction of feed that goes to the distillate (the D/F ratio);

  $D_{SP} = k\, F'$ = target for distillate flow;

  $B_{SP} = (1 - k)\, F' = F' - D_{SP}$ = target for bottoms flow.

- The value computed for the distillate flow target becomes the set point for the distillate flow controller.

- When the bottoms level is close to its target, the value computed for the bottoms flow becomes the set point for the bottoms flow controller. But as will be explained shortly, this is not the case when the control error in bottoms level is large.

5.4.1. Computing Product Flow Targets

In Figure 5.11, the targets for the distillate flow and bottoms flow are computed using the following equations:

$$D_{SP} = k\, F'$$
$$B_{SP} = F' - D_{SP}$$

Two simple function blocks are required to implement these equations:

Multiplier. This function block multiplies the corrected feed flow F' by the split factor k to obtain the target D_{SP} for the distillate flow.

Summer. This function block subtracts the distillate flow target D_{SP} from the corrected feed flow F' to obtain the target B_{SP} for the bottoms flow.

As Fractronic was originally presented, this logic was incorporated into a block designated as the "flow divider."

5.4.2. Feed Flow Correction Factor

The flow divider is based on the overall column material balance:

$$F = B + D.$$

Given a value for the split k, the distillate flow D and the bottoms flow B should be

$$D = k F,$$
$$B = (1-k) F = F - D.$$

This would work great in a perfect world. Furthermore, there would be no need for a bottoms level controller to close the overall material balance.

Unfortunately, the real world does not quite work this way. A small error accompanies every flow measurement. In the world of perfect measurements, the steady-state total material balance for the column would be

$$F - (B+D) = 0.$$

But in the real world, the result would be

$$F' - (B+D) = \varepsilon,$$

where ε is a small positive number or a small negative number.

The behavior depends on the sign of ε:

$\varepsilon > 0$. The feed flow exceeds the sum of the distillate flow and bottoms flow. The column slowly fills with liquid, which means that the bottoms level slowly increases.

$\varepsilon < 0$. The feed flow is less than the sum of the distillate flow and bottoms flow. The liquid is slowly being depleted from the column, which means that the bottoms level slowly decreases.

The question is how to force the column material balance to close in face of measurement errors in the three flows. It can be assumed that the error in the material balance closure is due solely to the measurement error in the feed flow (the largest of the three flows). For the measured value of the feed flow, the material balance does not close:

$$F - (D+B) = \varepsilon.$$

The desire is to obtain a corrected value F' for the feed flow for which the material balance closes:

$$F' - (D+B) = 0.$$

Figure 5.11 obtains the corrected value F' for the feed flow by multiplying the measured value F for the feed flow by the feed flow correction factor w. The question now is how to obtain a value for w. In Figure 5.11, the feed flow correction factor w is the output of the bottoms level controller. This controller behaves as follows:

• If $F' > (D + B)$, the values computed for D_{SP} and B_{SP} are too large. Too much material is being removed from the column, which causes the bottoms level to decrease. The bottoms level controller should decrease the value of w.

- If $F' < (D + B)$, the values computed for D_{SP} and B_{SP} are too small. Too little material is being removed from the column, which causes the bottoms level to increase. The bottoms level controller should increase the value of w.

Per this logic, the bottoms level controller should be direct acting—on an increase in the bottoms level, the bottoms level controller increases its output, which is the value of w.

5.4.3. Alternate Configuration for Bottoms Level Control

As Fractronic was originally presented, the P&I diagram contained a block designated the "F. F. Module" that provided alternate control logic for bottoms level depending on the deviation of the bottoms level from its target:

Magnitude of the bottoms level control error less than the tolerance. The logic in Figure 5.11 is used:

- The set point for the bottoms flow is the value computed by the flow divider.
- The bottoms level is controlled by adjusting the feed flow correction factor w. This loop responds very slowly, so it will not effectively respond to major upsets in the bottoms level.

Magnitude of the bottoms level control error exceeds the tolerance. The logic in Figure 5.12 is used. The bottoms level is controlled by manipulating the set point to the bottoms flow controller. This is a simple level to flow cascade.

For significant bottoms level control errors, fast corrective action is required. The bottoms level controller in Figure 5.12 would be tuned to respond aggressively to the control errors. There will be significant and rapid changes in the bottoms flow. This could have adverse consequences on the downstream processing equipment. However, excessive bottoms level control errors could lead to a shutdown on either high bottoms level or low bottoms level. Such shutdowns must be avoided if possible, even if it means upsetting the downstream units.

5.4.4. General Practice

In effect, Fractronic provides for both distillate-to-feed ratio and bottoms-to-feed ratio. This is not the general practice.

One of the product streams must be on level control. The general practice is to not ratio this product stream to the feed. The general practice is to provide

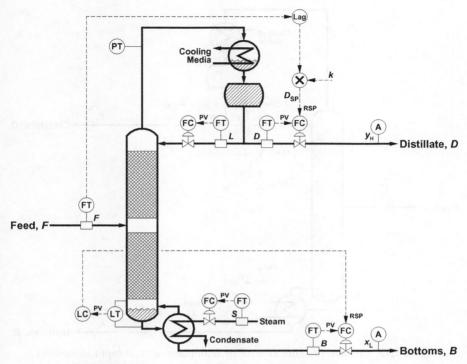

Figure 5.12. Level-to-flow cascade for bottoms level control.

ratios only for those product streams that are manipulated by a composition or temperature controller. The resulting configuration is illustrated in Figure 5.12. This is entirely consistent with the control configurations presented in the prior discussion regarding feedforward for feed flow changes.

In other words, the common practice is to install the configuration in Figure 5.12, often without the bottoms flow controller. Measuring the bottoms flow is sometimes difficult (high viscosity, high temperature, problems with buildups, etc.), so eliminating the need for this measurement can be appealing.

Why does Fractronic include a ratio for both product streams? In the article, the authors noted that on loss of the column feed pump, ratioing both product streams to the feed would quickly stop both product flows and switch the tower to total reflux. On loss of the feed to the tower for any reason, the product streams must be stopped quickly. With no feed flow, the product flows will quickly deplete the liquid within the tower. If too much liquid is removed, the bottoms level will be too low, causing a shutdown to be initiated on low bottoms level. However, there are other ways to address this problem; ratioing both product flows to the feed flow is not the common solution.

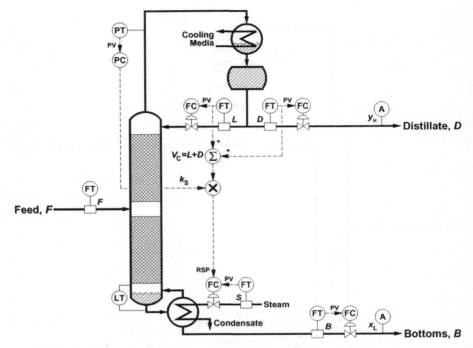

Figure 5.13. Column pressure control by manipulating heat input to reboiler.

5.5. FEEDFORWARD FOR COLUMN PRESSURE

The schematic in Figure 5.13 contains only those components of Fractronic that pertain to controlling the column pressure. Briefly, the approach is as follows:

- A summer computes the sum of the distillate flow D and the reflux flow L. At steady state, this would be the overhead vapor flow rate V_C from the top stage to the condenser.
- To produce a unit of overhead vapor, some number of units of steam is required. The factor k_S can be introduced as the unit of steam per unit of overhead vapor.
- The output of the summer can be considered to be the flow rate of the overhead vapor. How much steam is required to produce this overhead vapor? Multiply the overhead vapor flow (the output of the summer) by the steam-to-vapor factor k_S to obtain the set point for the steam flow controller.
- The value for the steam-to-vapor factor k_S is the output of the column pressure controller. More on this coefficient shortly.

5.5.1. Material Balance around the Vapor Space

The basis for the configuration in Figure 5.13 for controlling the tower pressure is a material balance around the column vapor space. Even in a simple two-product tower such as the one in Figure 5.13, this material balance has several terms:

- Vapor V_B produced by the reboiler. This is an input term.
- Vapor V_C condensed by the condenser. This is an output term.
- Vapor from the feed. If the feed is above its bubble point, this is an input term. If the feed is below its bubble point, this is an output term.
- Vapor from nonequimolal overflow. This could be either an input or output term, depending on the properties of the materials.
- Vapor associated with heat losses. Hopefully, this is a small term; otherwise, some investments in insulation should be recommended.

A simple relationship can be obtained by making certain assumptions:

- The feed enters at its bubble point (no significant change in vapor flow at the feed stage).
- The vapor flow within each separation section is constant (equimolal overflow).
- Negligible heat losses.

Under these assumptions, the steady-state material balance around the vapor space is as follows:

$$V_B - V_C = 0 \text{ or } V_C = V_B.$$

Under these assumptions, the steam flow S can be computed as follows:

$$S = k_S \, V_B = k_S \, V_C = k_S \, (D + L).$$

Furthermore, the value for the factor k_S can be computed from the latent heats of vaporization:

$$k_S = \frac{\lambda_V}{\lambda_S},$$

where

λ_V = latent heat of vaporization for vapor;
λ_S = latent heat of vaporization for steam.

5.5.2. Role of the Pressure Controller

In practice, the value of k_S cannot be computed as the ratio of the latent heats of vaporization. The feed to a column is rarely at its bubble point. Even materials such as light hydrocarbons exhibit some departure from ideality, so the assumption of equimolal overflow introduces some error. Heat losses should be small, but not zero. And finally, there are the ever-present measurement errors, which should also be small but not zero.

Probably the best source of a starting value for k_S is from the solution of the stage-by-stage separation model. Values for the distillate flow, the reflux flow, and the boilup can be obtained from this solution. The steam flow can be computed from the boilup using the ratio of latent heats. The value of k_S is the steam flow divided by the sum of the distillate and reflux flows.

If the value of k_S is not correct, the column pressure will be affected as follows:

- **Value for k_S too large.** The steam flow is too high, so the boilup is too high. The column pressure will be increasing. The tower pressure controller should respond by decreasing the value of k_S.
- **Value for k_S too small.** The steam flow is too low, so the boilup is too low. The column pressure will be decreasing. The tower pressure controller should respond by increasing the value of k_S.

The pressure controller should be reverse acting. On an increase in the column pressure, the pressure controller should decrease its output, which is the value of k_S.

5.5.3. Steam-to-Feed Ratio

In the previous discussion pertaining to feedforward for feed flow changes, a steam-to-feed ratio was recommended. The configuration in Figure 5.13 does not appear to provide such a ratio.

The pressure control loop alone does not provide such a ratio. However, one must also consider the contribution of the other loops within the control configuration. Specifically, the contribution of the following two loops together must be considered:

Column material balance (Fig. 5.11 or 5.12). Both configurations provide a ratio of the distillate flow D to the feed flow F. A unit change in the feed flow causes the distillate flow to change by k units.

Material balance for the vapor space (Fig. 5.13). A unit change in the distillate flow D causes the steam flow to change by k_S units.

The combined effect is that a unit change in the feed flow causes the steam flow to change by $k\,k_S$ units. When the control configuration in Figure 5.13 is

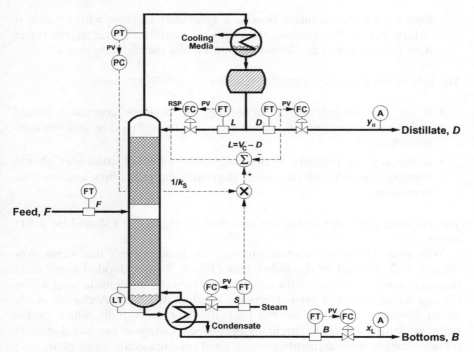

Figure 5.14. Column pressure control by manipulating reflux flow.

combined with the control configuration in either Figure 5.11 or 5.12, the net result is that the steam flow is ratioed to the feed flow.

5.5.4. Manipulating Reflux to Control Pressure

Controlling the column pressure by manipulating the heat to the reboiler is unusual, despite the fact that it usually provides good pressure control. A problem arises when the bottoms level and the bottoms composition must be controlled. This will be discussed shortly.

The pressure control configuration in Figure 5.14 controls the column pressure by manipulating the reflux flow. In a flooded condenser, changes in the reflux flow lead to changes in the condensation rate and the heat transfer rate in the condenser. The basis for the configuration in Figure 5.14 is as follows:

- Dividing the steam flow S by the factor k_S (or multiplying by $1/k_S$) gives the overhead vapor flow V_C, which is the vapor condensation rate within the condenser.
- The vapor condensed in the condenser must be removed as either distillate flow or returned to the column as reflux flow. In the configuration in

Figure 5.14, the distillate flow is a measured variable whose value is determined by the distillate composition controller. Therefore, the reflux flow L is the vapor condensation rate V_C less the distillate flow D.

The role of the pressure controller remains essentially the same:

- If the column pressure is increasing, the pressure controller should increase the value of $1/k_S$, which increases the reflux flow and the condensation rate.
- If the column pressure is decreasing, the pressure controller should decrease the value of $1/k_S$, which decreases the reflux flow and the condensation rate.

The pressure controller in the configuration in Figure 5.14 should be direct acting.

Why does Fractronic control pressure by manipulating the steam flow (Figure 5.13) instead of the reflux flow (Figure 5.14)? Flooded condensers respond slowly—a change in the reflux flow affects the condensate level within the condenser, which in turn affects the condensation rate. A change in the steam flow will affect the column pressure far more rapidly, which means better pressure control. As originally proposed, Fractronic was based entirely on temperature measurements, making good pressure control essential.

5.6. PRODUCT COMPOSITIONS

In most double-end composition control applications, the most appropriate approach is to control one of the compositions by manipulating a term in the material balance and the other composition by manipulating a term in the energy balance. The Fractronic configuration in Figure 5.10 does exactly this:

Distillate composition. Control by manipulating the product draws.

Bottoms composition. Control by manipulating the reflux flow.

5.6.1. Distillate Composition

Figure 5.15 is the configuration in Figure 5.11 with the addition of a controller for the distillate composition. The output of the distillate composition controller is the split factor k, which is the input to the function blocks that compute the distillate flow target and the bottoms flow target from the feed flow.

This approach to controlling distillate composition is commonly applied, especially to columns with a variable feed flow rate. The only unusual part of Figure 5.15 is that both the distillate and the bottoms flow are computed from the feed flow and the split factor. The more customary configuration is obtained by adding the distillate composition controller to the control configuration in

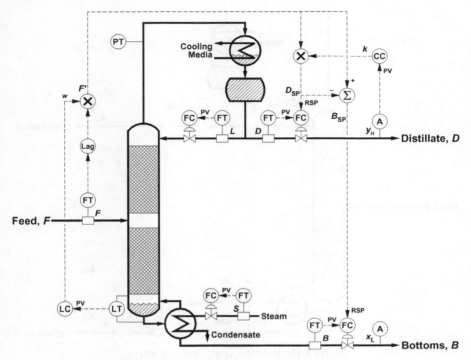

Figure 5.15. Distillate composition control.

Figure 5.12, which computes only the target for the distillate flow from the feed flow and the split factor.

If the composition of the heavy key in the distillate product is increasing, too much distillate product is being removed. The distillate composition controller should decrease the split factor k. The composition controller must be reverse acting—on an increase in the composition of the heavy key in the distillate product, the controller should decrease its output, which is the split factor k.

5.6.2. Bottoms Composition

Figure 5.16 presents the control configuration for the bottoms composition. The output of the bottoms composition controller adjusts the set point to the reflux flow controller. This is a composition-to-flow cascade, which is in keeping with the customary recommendation that slow loops such as composition should output to a flow controller instead of outputting directly to a control valve.

Objections are likely to be raised regarding the configuration in Figure 5.16. With regard to level and composition controls, the conventional wisdom is to control each variable by manipulating a valve opening or flow set point on the

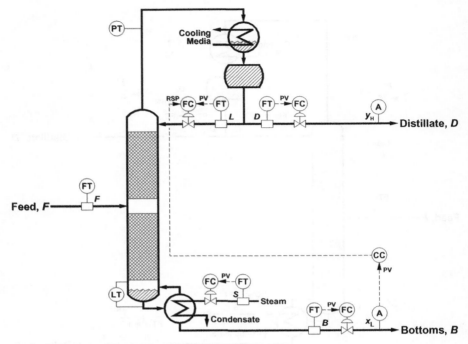

Figure 5.16. Bottoms composition control.

same end of the tower. With this logic, there are two options for controlling the bottoms composition:

- Manipulate the heat to the reboiler (which determines the boilup).
- Manipulate the bottoms flow.

However, neither is available. The bottoms flow is being manipulated to control the bottoms level. The heat to the reboiler is being manipulated to control the column pressure. This is a consequence of controlling column pressure with boilup—no manipulated variable is available at the bottom of the tower that can be used to control bottoms composition.

5.6.3. Action for Bottoms Composition Controller

Based on the control configuration in Figure 5.16, the expected effect of increasing the reflux flow is to increase the amount of the light components in the bottoms. If the composition of the light key in the bottoms product is increasing, the bottoms composition controller should decrease the reflux flow—that is, the controller should be reverse acting.

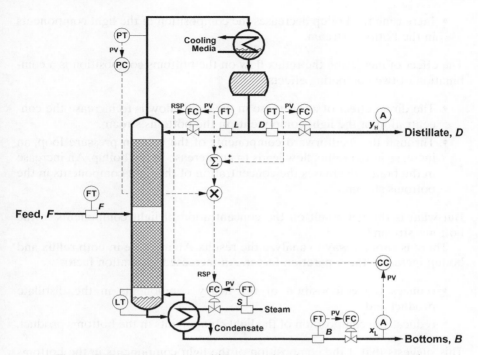

Figure 5.17. Bottoms composition control and tower pressure control.

But for the Fractronic configuration, this is not correct. In any control configuration that includes feedforward components, their impact on cause-and-effect relationships must also be considered. That is, the analysis must encompass two considerations:

- The direct effect of the change in the manipulated variable.
- The effect of any feedforward actions that result from the change in the manipulated variable.

Figure 5.17 illustrates the tower pressure control configuration along with the composition-to-flow cascade for controlling the bottoms composition. What is the contribution of the feedforward elements for an increase in the reflux flow?

- The increase in reflux flow increases the sum of the reflux flow plus distillate flow.
- The sum of the distillate flow and reflux flow is multiplied by the steam-to-vapor ratio to obtain the set point for the reboiler steam flow controller. Increasing the heat to the reboiler increases the boilup.

- Increasing the boilup decreases the composition of the light components in the bottoms stream.

The effect of increasing the reflux flow on the bottoms composition is a combination of two competing effects:

- The direct effect of an increase in the reflux flow is to increase the concentration of the light components in the bottoms stream.
- Through the feedforward components of the column pressure loop, an increase in the reflux flow leads to an increase in the boilup. An increase in the boilup decreases the concentration of the light components in the bottoms stream.

But what is the net result on the concentration of light components in the bottoms stream?

There is another way to analyze the results. An increase in both reflux and boilup increases the separation factor. An increased separation factor

- reduces the composition of the heavy components in the distillate product and
- reduces the composition of the light components in the bottoms product.

This suggests that if the composition of the light components in the bottoms product is increasing, the separation factor must be increased. The bottoms composition controller should increase the reflux flow, which means a direct acting controller. The feedforward components of the pressure loop will increase the boilup. Together these increase the separation factor.

This logic is correct provided the feedforward elements within the tower pressure loop are functioning. But what if this feedforward logic must be disabled for some reason (such as a faulty measurement device)? The bottoms composition control configuration becomes the composition-to-flow cascade in Figure 5.16. For this configuration, the controller action should be reverse and not direct.

When the feedforward logic is based on basic measurements (flow, pressure, etc.), disabling the feedforward logic should be an infrequent occurrence. However, logic should be incorporated into the control configuration in Figure 5.17 to force the bottoms composition controller to manual should the steam flow controller not be using its remote set point input. It is not necessary that the tower pressure controller be on automatic. The tower pressure controller is providing the feedback trim and is not part of the feedforward elements.

REFERENCE

1 Wright, R. M. and A. W. Johncock, A Better Approach to Distillation Control, Instruments and Control Systems, June 1976.

6

UNIT OPTIMIZATION

With very few exceptions, the optimum operating point for a distillation column is at a constraint. How close one can operate to the constraint depends on the variance in the process variable to which the constraint pertains. Consequently, one route to enhancing process performance involves two steps:

- Narrow the variance.
- Shift the target.

But for distillation columns (and many other processes), there is a twist that is too often ignored: As the target is shifted toward the specification limit, the variance increases. This places increased emphasis on control capabilities—you cannot optimize something that you cannot control.

When shifting the target for a product composition, the potential benefits include the following:

- Reduced energy consumption.
- Improved recovery.
- Increased throughput.
- Maximize low value impurity in a high value product stream.

Starting with the first energy crisis in the 1970s, the emphasis in distillation has been on energy conservation. But except where one of the utilities is expensive

Distillation Control: An Engineering Perspective, First Edition. Cecil L. Smith.
© 2012 John Wiley & Sons, Inc. Published 2012 by John Wiley & Sons, Inc.

(such as refrigerant), the economic returns from the other possibilities will likely exceed that of energy conservation.

Improved recovery should always be explored. For a given feed rate, is it possible to adjust conditions within the tower to increase the flow rate of the desired product? If so, the return always exceeds the return from energy conservation. Furthermore, this can produce benefits in associated unit operations, especially where the distillation column is within a recycle loop.

Increased throughput gives the highest returns but is the least likely to be possible. Increased throughput is only beneficial if the distillation tower is limiting plant throughput. Where some other unit operation is limiting the plant throughput, increasing the throughput of the distillation tower is of no interest.

The impurities in a product are sold at the price of the product. If the value of the impurity is less than the value of the product, the amount of that impurity in the product should be as much as the specifications permit. However, this is usually worth pursuing only in large production units.

6.1. ENERGY AND SEPARATION

Consider a distillation column that is operated with fixed values for

- feed flow, feed composition, and feed enthalpy;
- the distillate recovery (D/F);
- the column pressure.

The distillate composition, measured as the concentration of the heavy key (an impurity), must meet or exceed specifications.

The steam flow to the reboiler (which determines the boilup) is at the discretion of the process operator. The boilup determines the distillate composition, or vice versa, obtaining a specified distillate composition requires a certain boilup. The operating target for the distillate composition determines the energy required to operate the tower.

To decrease the energy utilization in such a tower, there is only one option: adjust the target for the distillate composition. Shifting the target to a value closer to the product specifications reduces the energy required in the tower. But for reasons that will be explained shortly, process operators prefer to operate towers with a higher than required target for the distillate composition, which gives a product with a higher purity than required by the specifications.

6.1.1. Column Operating Line

This graph relates the distillate composition and the boilup (or something related to the boilup, such as the steam flow to the reboiler). Figure 6.1

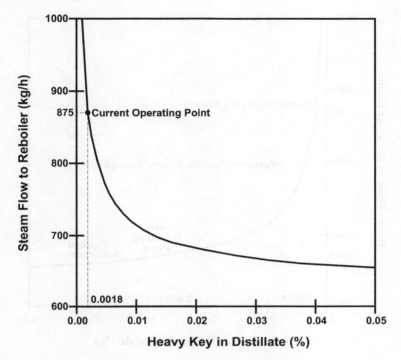

Figure 6.1. Column operating line.

presents the column operating line for the depropanizer presented in Section 1.9.

The column operating line is normally constructed using the following approach:

1. Collect data from the tower at its current operating point and calibrate the stage-by-stage model to the column at this point. This gives one point on the operating line, which is so designated in Figure 6.1.
2. For other values of the boilup (or steam flow to the reboiler), compute the corresponding values for the distillate composition. Plotting these gives the graph in Figure 6.1.

The graph in Figure 6.1 is typical of distillation columns. High purities in a product can only be attained by consuming large amounts of the utilities. If a column is currently producing a product that greatly exceeds the required purity, considerable energy savings are possible.

Prior to the first energy crisis in the 1970s, operating columns to produce excessively pure products was the norm. Except for columns requiring an expensive utility such as refrigerant, energy was basically viewed as "free" and

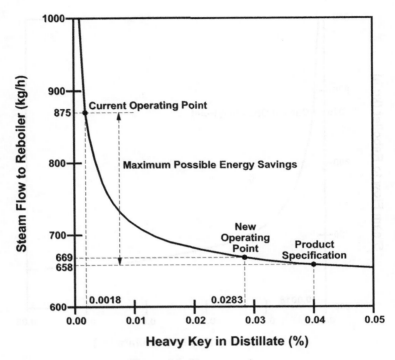

Figure 6.2. Energy savings.

was consumed with no regard to either cost or availability. The energy crisis changed the ground rules, and operating practices were revised accordingly.

6.1.2. Potential Energy Savings

The maximum allowable value for the heavy key in the distillate is determined by the product specifications. The graph in Figure 6.2 contains a point designated "product specification," which is 0.04% heavy key in the distillate. This establishes the maximum possible energy savings:

Steam flow to reboiler at current operating point: 875 kg/h
Steam flow to reboiler at product specification: 658 kg/h
Maximum possible energy savings: 217 kg/h

This is a potential reduction of 25% in the energy consumption! In the 1970s, such cases were common, but few remain today.

One cannot operate a column exactly at the product specification. How much operating margin is required depends on various factors, one being the frequency and magnitude of the upsets to the tower. Fortunately, the shape of the operating line in Figure 6.2 permits the bulk of the energy savings to be realized without operating extremely close to the product specification.

Figure 6.2 contains a point designated "new operating point," which is 0.0283% heavy key in the distillate (this point corresponds to the data in Fig. 1.14). The actual energy savings are as follows:

Steam flow to reboiler at current operating point: 875 kg/h

Steam flow to reboiler at new operating point: 669 kg/h

Actual energy savings: 206 kg/h

Even though the column is not being operated very close to the specification limit, 95% of the maximum possible energy savings is realized.

When a column is being operated so as to produce an excessively pure product, significant energy savings will be possible. But if the degree of over-separation is modest, the potential energy savings are unlikely to justify any effort beyond simply bringing this to the attention of production personnel. Using the example in Figure 6.2, the new operating point of 0.0283% heavy key in the distillate is somewhat conservative given that the product specification is 0.04% heavy key in the distillate. However, the potential energy savings is a mere 11 kg/h of steam.

6.1.3. Variance

From historical data on column operating conditions, the following two parameters can be computed for the distillate composition (the heavy key in the distillate):

Mean. This value is the "current operating point" in Figure 6.2.

Variance. This can be characterized by its standard deviation σ.

Adjusting the target to a value closer to the specification limit seems like a "no-brainer" that could be approached as follows:

- Select some "operating margin" such as a 2σ or 3σ for the variance.
- Shift the target as close to the specifications limit as the "operating margin" permits.

However, this logic assumes that shifting the target has no effect on the variance. For distillation, this is not the case.

Most of the early attempts at energy conservation were not prepared for this. Columns that operated smoothly ($\sigma \cong 0$) with excess separation performed very differently when the target was moved closer to the specification limit. One of the best articles [1] to explain this came from the Applied

Automation organization, which at that time was a wholly owned subsidiary of Phillips Petroleum.

For distillation columns, shifting the target has a significant impact on the variance. Shifting the target in the direction of the specification limit causes the variance to increase. The target cannot be simply shifted. Simple controls are adequate for operating with an excessively pure target, but not for operating with a target close to the specification limit. The column controls must be upgraded, which can only be undertaken provided the costs are offset by the reduction in energy consumption.

6.1.4. Shape of the Operating Line

The column operating line in Figure 6.2 reflects the sensitivity of the distillate composition (the heavy key in the distillate) to changes in the boilup. The operating line is concave up with the following consequences:

Original operating point. The operating line exhibits a steep slope, so disturbances in boilup are attenuated. The process sensitivity is the sensitivity of the distillate composition to changes in the boilup, which is the reciprocal of the slope of the operating line in Figure 6.2. At the original operating point, the process sensitivity is low.

New operating point. The operating line does not exhibit a steep slope, so disturbances will not be attenuated. The process sensitivity is much higher, causing disturbances in the boilup to have a much larger effect on the distillate composition.

For most processes, the operating line is concave up, so shifting the target will degrade the performance of the controls. The process is more sensitive to disturbances, and this is reflected in the performance of the controls.

A few applications have a linear operating line. Filling systems (e.g., filling jars with mayonnaise) have a linear operating line. Shifting the target has no effect on control performance. If the operating line is concave down, shifting the target closer to the specification limit would make the process easier to control. There seem to be very few of these, and perhaps none.

6.1.5. Disturbances

One of the bases for the operating line in Figure 6.2 is a constant boilup, which is achieved by approaches such as maintaining a constant steam flow to the reboiler. However, a small variance is likely present in the boilup.

For both the original operating point and the new operating point, Figure 6.3 illustrates the effect of the same variance in the boilup on the impurities in the distillate. The variance is the result of disturbances in the boilup, which should not be affected by shifting the target. In Figure 6.3, the variance in boilup is the same at both operating points. However, the variance in the

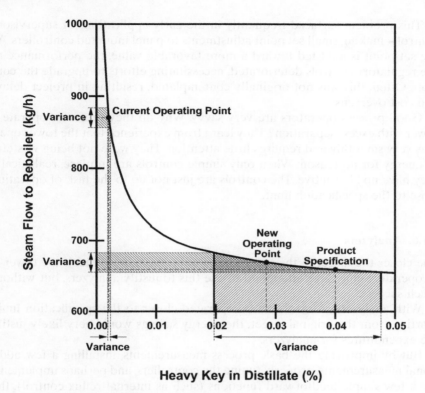

Figure 6.3. Propagation of variance.

impurities in the distillate increase dramatically as the target is shifted toward the specification limit.

Figure 6.3 is specifically constructed for disturbances in boilup. Distillation towers are subjected to a variety of upsets (feed composition, feed enthalpy, cooling water temperatures, etc.). What is the effect of these other disturbances? Any disturbance to a tower can be represented as an equivalent disturbance in the boilup. For example, a change in the feed composition will have a certain effect on the distillation composition. There is some equivalent change in the boilup that will have the same effect. Consequently, the variance in the boilup illustrated in Figure 6.3 is not just the actual variance in boilup, but is the variance in boilup that is equivalent to the effect of all of the disturbances to the tower.

Although this analysis is largely qualitative, it is consistent with observations in the field—shifting the target toward the specification limit somehow impacts the performance of the existing controls in a noticeably negative manner. Controls deemed to provide adequate performance were found to be lacking.

This problem surfaced frequently in the early applications of supervisory control—making small set point adjustments to panel mounted controllers. As the set point is adjusted toward a more favorable value, the performance of the regulatory controls deteriorated, necessitating efforts to upgrade the controls. Often, this was not originally contemplated, resulting in project delays and cost overruns.

Good process operators are very savvy. Why do they prefer to operate a tower with excess separation? They learn from experience that the tower operates very smoothly and requires little attention. They were not being wasteful of energy for no reason. When only simple controls are in place, realistically they have no alternative. The controls are just not up to the task of operating close to the specification limit.

6.1.6. Analyzers

The closer the target to the specification limit, the more difficult the tower is to operate. Some have attempted to use this to justify analyzers, but without much success.

With analyzers, a tower can be operated closer to the specification limit. Starting from the original target, the energy savings would very likely justify the expenditures for analyzers.

But by improving the basic process measurements, installing a few additional measurements, properly tuning the controllers, and perhaps implementing a few simple feedforward functions (such as internal reflux control), the target can be shifted in the direction of the specification limit. With the concave up shape of the column operating line, more energy is conserved with the initial shifts of the target than with the same shift when closer to the specification limit. Generally, one is able to achieve significant reductions in energy for a relatively (compared with analyzers) nominal expenditure of time and money. With analyzers, one could go further. However, the remaining potential energy savings are rarely sufficient to justify the expenditures for analyzers.

6.2. OPTIMIZATION OF A COLUMN

Various aspects of optimization as applied to an individual distillation column will be presented in the context of the column in Figure 6.4, which is very similar to one used by Baxley [2]. The key aspects are as follows:

- The feed flow, feed composition, and feed enthalpy are determined by the upstream unit operations.
- The condenser is a partial condenser, with refrigerant as the cooling media.

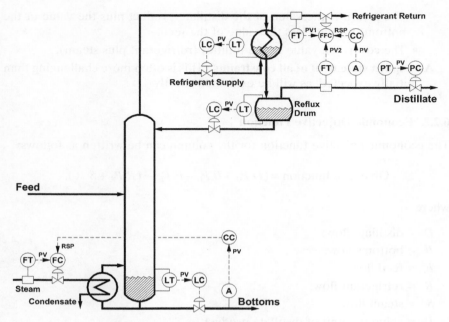

Figure 6.4. Control configuration for column.

- Column pressure is controlled by the valve on the distillate vapor product.
- The distillate composition is controlled by manipulating the set point of a ratio or flow-to-flow controller that maintains the specified ratio of refrigerant flow to distillate flow. In essence, the manipulated variable for the distillate composition is the reflux to distillate (L/D) ratio. The distillate product is a salable product and must meet tight specifications.
- The bottoms composition is controlled by manipulating the steam to the reboiler, which determines the boilup.

Only the measurements required by the control loops are illustrated in Figure 6.4. In what follows, it is assumed that additional measurements are either available or can be installed.

6.2.1. Formulation

The objective is to optimize the recovery of the distillate product, which is a salable product. The formulation of the optimization problem requires the following:

A clear objective function. Economic objective functions are simply the returns less the costs. For the distillation column, these are as follows:

- The return is the value of the distillate product plus the value of the bottoms product less the value of the feed.
- The cost is the value of the utilities (refrigerant plus steam).

An explicit statement of all constraints. This is often more challenging than initially expected, as will be examined shortly.

6.2.2. Economic Objective Function

The economic objective function for the column can be written as follows:

$$\text{Objective function} = (D\,P_D + B\,P_B - F\,P_F) - (R\,P_R + S\,P_S),$$

where

D = distillate flow;
B = bottoms flow;
F = feed flow;
R = refrigerant flow;
S = steam flow;
P_D = value per unit of distillate product;
P_B = value per unit of bottoms product;
P_F = value per unit of feed;
P_R = value per unit of refrigerant;
P_S = value per unit of steam.

The term for the feed flow is constant, so it has no effect on the optimum. If desired, the feed term can be omitted from the formulation of the optimization logic for the column.

Most production facilities have established numbers for the values of the utilities. However, this is probably not the case for the feed and one or more product streams.

6.2.3. Product Values

Obtaining a reasonable number for the value of a product from a tower within a production facility can be a challenge. The possible situations include the following:

- Where the product is a salable product, market pricing is available. But for optimization purposes, this value should be adjusted for factors such as the cost of sales.
- Where the product is transferred to another business unit within the company, a transfer cost is available. Establishing the transfer cost often involves company politics. Optimization requires a realistic number, not

a political number. If a plant-wide or corporate optimization effort encompasses both the producer and the consumer, a realistic value can be derived from these results.

- Accounting may have a number that it uses for the product value within its reporting systems. However, accounting's focus is entirely on the bottom line. In many cases, the product value is used to compute a cost on one balance sheet, and then used to compute a return on another balance sheet. When the balance sheets are combined, the cost cancels the return, leaving no effect on the bottom line. From accounting's perspective, any number will do.

- The product may be an internal plant stream that is produced by one unit operation and consumed by the next. It is quite possible that no value has ever been assigned to this product. If a plant-wide optimization effort is in place, this is the best source of a realistic value.

6.2.4. Incremental Formulation

When optimizing an individual tower, the point of reference should always be the current column operating conditions. Measurements are available for the product flows, the utility flows, the compositions, and so on. Optimization can be implemented by making changes that are in the direction of the optimum.

Let ΔD be the proposed change in the distillate flow. If the distillate flow increases by ΔD, then the bottoms flow decreases by ΔD. The incremental return and the incremental cost are given by the following equations:

Incremental return = $\Delta D \, (P_D - P_B)$

Incremental cost = $\Delta R \, P_R + \Delta S \, P_S$

where ΔR and ΔS are the incremental utilities required to increase the distillate product by ΔD. Optimization involves changing D in the direction for which the incremental return exceeds the incremental cost. Calculating both presents challenges:

Incremental return. Realistic numbers are required for the product values.

Incremental cost. The incremental utilities ΔR and ΔS required to achieve ΔD must be computed.

6.2.5. Incremental Utilities

Today, all distillation calculations are based on the stage-by-stage separation model. This model relates actual values; that is, it relates the distillate flow

D and the energy flows, from which the utility flows R and S can be calculated. To use such a model to calculate incremental changes requires the following steps:

1. Obtain data for current column operating conditions (flows, compositions, etc.).
2. Calibrate the stage-by-stage separation model to the current column operating conditions.
3. Compute a new steady-state solution as follows:
   - Change the distillate flow by ΔD.
   - Determine the boilup that gives the same distillate composition as for the current column operating conditions.
   - From the energy flows for this solution, compute ΔR and ΔS.

This is a lot of work.

Reformulating the process model to relate incremental changes in various variables greatly reduces the work. If the incremental changes are restricted to small changes, high accuracy is not necessary. But in distillation, the total focus has been on developing stage-by-stage models for design, troubleshooting, and so on. Reformulating to incremental models has not been a high priority.

6.2.6. Noneconomic Objective Functions

Examples of noneconomic objective functions include the following:

- Maximum production of a product (e.g., the distillate product from the column in Fig. 6.4).
- Minimum energy (e.g., minimize refrigerant consumption for the column in Fig. 6.4).
- Minimum effluent rate.
- Maximize the amount of a low value impurity in a high value product.

A major incentive is that product values are not required. The solution is to operate at one of the constraints.

However, one must be very careful. For the column in Figure 6.4, is it safe to assume that it is always economically attractive to produce more distillate product? Producing more distillate product means reducing the light components in the bottoms. As these compositions become small, the energy flows increase very rapidly. When one of the utilities is expensive, the cost of the additional energy to produce a unit of distillate product might very well exceed the value of the incremental product.

6.3. CONSTRAINTS IN DISTILLATION COLUMNS

In formulating any optimization problem, obtaining explicit statements for all of the constraints is crucial. The solution of the optimization problem is usually to operate as close as possible to one of the constraints. But if that constraint is not included in the formulation, the optimization logic will attempt to operate in violation of the constraint, usually by taking control actions opposite to the appropriate control actions.

No technology is available that can assure that all constraints have been identified. Identifying constraints rests entirely on one's understanding of the process. Fortunately, distillation is a common unit operation, so considerable experience is available. The constraints generally fall into the following categories:

- Tower internals.
- Condenser.
- Reboiler.
- Temperatures.
- Control valves.

6.3.1. Tower Internals

There are three constraints associated with the tower internals:

Tower flooding. This basically imposes an upper limit on the internal vapor flows, notably the boilup (or heat input to the reboiler). This constraint applies to both tray and packed towers.

Minimum vapor velocity. This constraint only applies to tray towers. If the vapor velocity is too low, the weeping of liquid through the perforations in the trays becomes excessive, possibly to the point of draining all liquid from the trays.

Minimum liquid flow. This constraint only applies to packed towers. The packing must be completely wet at all times. This requires a minimum reflux flow to the tower.

Can some constraints be omitted from in the formulation? If the objective is to maximize the production of distillate product, tower flooding could potentially be the limiting constraint. The minimum vapor velocity or minimum liquid flow should never be the limiting constraint. But if the objective is to minimize the use of refrigerant, tower flooding should not arise, but the constraints on minimum vapor velocity or minimum liquid flow could be encountered. Omitting one or more constraints simplifies the formulation, but at some peril—should an omitted constraint be encountered during column operations, the optimization logic will recommend a control action opposite to what is appropriate.

6.3.2. Condenser

The formulation of the constraint depends on what limits the heat transfer in the condenser. As noted previously, a heat transfer process may be operating in either the heat transfer limited mode or the media limited mode. The constraints are different for the two modes:

Heat transfer limited. For most exchangers, this constraint determines the maximum heat transfer. For the refrigerant condenser in Figure 6.4, the fraction utilization at the current operating conditions is

$$\text{Fraction utilization} = \frac{T_R - T_C}{T_{R,MIN} - T_C},$$

where

T_C = temperature of the condensing vapor;
T_R = current temperature of the vaporizing refrigerant;
$T_{R,MIN}$ = minimum temperature attainable by the vaporizing refrigerant.

Media limited. There are two possibilities:

- The control valve determines the upper limit on the refrigerant flow. As oversized valves are the norm, this is unlikely.
- The refrigerant plant has limited capacity, and refrigerant is allocated to each user of the refrigerant. No user may exceed the allocated amount. Usually, this is enforced based on a period average (hourly average will be used herein). Short duration excursions above the allocated amount are usually tolerated.

6.3.3. Reboiler

The formulation of the constraint depends on what limits the heat transfer in the reboiler. Being a heat transfer process, the reboiler may be operating in either the heat transfer limited mode or the media limited mode. The constraints are different for the two modes:

Heat transfer limited. For most reboilers, this constraint determines the maximum heat transfer. For the steam heated reboiler illustrated in Figure 6.4, the fraction utilization at the current operating conditions is

$$\text{Fraction utilization} = \frac{T_S - T_B}{T_{S,SAT} - T_B},$$

where

T_B = temperature of the boiling liquid within the reboiler;

T_S = current temperature of the condensing steam within the reboiler;

$T_{S,SAT}$ = saturation temperature at the steam supply pressure.

Media limited. There are two possibilities:

- The control valve determines the upper limit on the steam flow. As oversized valves are the norm, this is unlikely.
- The steam plant has limited capacity, and steam is allocated to each user. Allocation procedures for steam are rare, so this is also unlikely.

6.3.4. Temperatures

Probably the two most common examples of constraints on temperatures are the following:

Bottoms temperature. Some products decompose or otherwise degrade at elevated temperatures. The highest temperature in a column is in the reboiler. Where temperature has an adverse affect on the product, an upper limit may be imposed on the bottoms temperature.

Overhead temperature. Cryogenic towers operate at low temperatures, sometimes so low that issues arise with regard to the metal used to fabricate the tower. At very low temperatures, metals become brittle and prone to failure. Where such issues arise, a minimum limit may be imposed on the overhead temperature.

6.3.5. Control Valves

Potentially there are two constraints associated with each control valve:

Valve fully closed. In towers, driving control valves fully closed is unusual. However, a major exception is hot gas bypass valves for which maximum cooling occurs when the valve is fully closed.

Valve fully open. Encountering such a constraint is definitely a possibility. But with the common practice of oversizing valves and other aspects of the process, other issues often arise before the valve is fully open. The most common is probably heat transfer processes where the transition from media limited to heat transfer limited usually occurs before the valve is fully open.

The most probable situation where a control valve limit is encountered is when a column is being operated under conditions other than those for which it was designed. The following is an example (a.k.a. war story) from oil refining. Switching a refinery from a light crude to a heavy crude significantly increases the heavy components in the feeds to some towers, resulting in a higher and more viscous bottoms flow. Will the pumping capacity for the bottoms be adequate? When the answer is "marginal," the temptation is to go with what is currently installed. But a potential result is that at times the pumping capacity is inadequate, which causes the bottoms level controller to drive the bottoms flow valve fully open. The resulting bottoms flow is insufficient, and the level in the bottoms continues to increase (in essence, tower begins to fill with liquid).

Such situations tend to be temporary. Once the problem is recognized, efforts are initiated to upgrade the pumping capacity. Such constraints can be incorporated into the optimization logic, but is it necessary to expend effort on a temporary problem? Depends on what is meant by "temporary"—the next turnaround could be several months away.

6.4. CONTROL CONFIGURATIONS FOR SINGLE CONSTRAINT

The objective is to produce as much distillate product as possible subject to the following three constraints:

- Heat transfer limit at the reboiler.
- Tower flooding. Although not shown in Figure 6.4, a measurement of tower pressure drop is available.
- Refrigerant allocation at the condenser. This tower is not permitted to consume more refrigerant that is allocated to it.

In this section, it will be assumed that the limiting constraint can be safely identified. For each of the above three constraints, control configurations will be proposed to operate at that constraint. Applications where all three constraints must be incorporated into the control logic are the subject of the next section.

6.4.1. Degrees of Freedom

The control configuration in Figure 6.4 provides double-end composition control. For a specified target for the distillate composition and a specified target for the bottoms composition, there is only one solution. The number of degrees of freedom is zero.

As the distillate product is salable, the target for the composition of the distillate product must be maintained at all times. To pursue optimization, there are two options:

Disable the bottoms composition controller, which permits the optimization logic to manipulate the steam flow to the reboiler. The bottoms composition becomes a dependent variable whose value changes as the optimization logic makes its adjustments.

Permit the optimization logic to adjust the set point to the bottoms composition controller. The bottoms composition is controlled, but not to a fixed target.

The net result of these two approaches is basically the same—the bottoms composition is said to "float."

6.4.2. Individual Constraints

For each of the three constraints listed above, two control configurations will be presented on the basis that only this constraint must be considered:

- A configuration that does not retain the bottoms composition controller. These configurations are usually simpler.
- A configuration that retains the bottoms composition controller. The resulting configuration is usually a cascade. Optimization is normally executed on a slow time frame, even slower than composition or temperature controls. However, concerns often arise regarding the dynamic separation between the loops in the cascade.

Another issue often arises. Some control configuration is currently installed in the production facility. Plant personnel are familiar with it, including its limitations and quirks. Changes to this configuration will certainly require some discussions. The path of least resistance is to layer optimization on top of the current control configuration with as few changes as possible.

6.4.3. Reboiler Heat Transfer

There is a simple way to obtain the maximum heat transfer rate from the reboiler: switch the bottoms composition controller and steam flow controller to manual and fully open the steam valve.

The disturbances to the boilup come from two sources:

Steam supply. Without the steam flow controller, disturbances such as steam supply pressure will affect the boilup.

Process side. Any change in the bottoms temperature affects the ΔT in the reboiler and, consequently, the heat transfer rate and the boilup. Possible upsets to bottoms temperature include the following:

- Column pressure. The pressure controller should maintain constant column pressure, but any excursions in column pressure will be compounded by changes in the boilup.
- Off-key components in the feed.

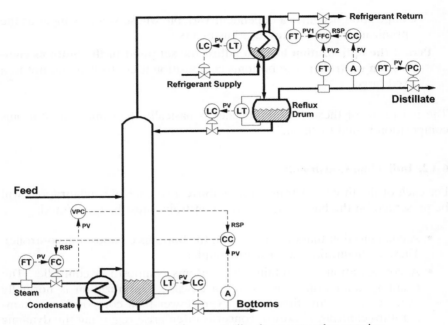

Figure 6.5. Valve position controller for steam valve opening.

Regardless of the disturbance, the reboiler always delivers the maximum possible heat input to the tower. However, the column is now exposed to a number of disturbances.

How can the bottoms composition controller be retained but still operate the reboiler close to its maximum heat transfer capability? One approach is to adjust the set point to the bottoms composition controller such that its output is a desired value, such as 90% open.

The valve position controller (VPC) configuration in Figure 6.5 is one approach. The output of the steam flow controller (the reboiler steam valve opening) is the measured variable for the VPC. The VPC adjusts the set point for the bottoms composition controller until the steam valve opening is equal to the VPC's set point. A typical value for this set point is 90% or 95%. If the steam valve opening is increasing, the VPC increases the bottoms composition controller set point, which is the target for the light key (an impurity) in the bottoms.

VPCs are commonly used in the fashion as illustrated in Figure 6.5. However, this configuration will perform satisfactorily only if the exchanger is media limited. But most reboilers are heat transfer limited. If so, a steam valve opening of 90% is in the heat transfer limited region where the steam valve opening has little effect on the heat transfer. The bottoms composition controller will not function properly in this region.

For a steam-heated reboiler, the fraction of the heat transfer capability that is being utilized must be calculated from the following equation:

$$\text{Fraction utilization} = \frac{T_S - T_B}{T_{S,SAT} - T_B},$$

where

T_B = temperature of the boiling liquid within the reboiler;

T_S = current temperature of the condensing steam within the reboiler;

$T_{S,SAT}$ = saturation temperature at the steam supply pressure.

This can be implemented in two ways:

Computed process variable (PV) for VPC. The fraction is computed from the current measurements of bottoms temperature, condensing steam temperature, and steam saturation temperature. This computed value is the PV for the VPC. A reasonable target for the fraction utilization is 90%.

Computed set point (SP) for VPC. The measured variable input to the VPC is the condensing steam temperature. The desired value for the condensing steam temperature is computed from the bottoms temperature, the steam saturation temperature, and the desired value for the fraction utilization (e.g., 90%). The computed value for the condensing steam temperature is the set point for the VPC.

Figure 6.6 presents a simpler implementation of the latter option that is possible when only small variations occur in the bottoms temperature and the steam supply pressure. The computation for the VPC set point is performed "off-line" and the result used as the set point for the VPC. Since the PV is the condensing steam temperature, the controller would customarily be designated TC (temperature controller) instead of VPC. However, VPC more accurately reflects the role of this controller and is used in Figure 6.6.

6.4.4. Tower Flooding

If the tower is properly designed and if the tower is operating under the conditions for which it was designed, tower operations should be restricted by the constraint associated with the most expensive part of the equipment. For towers, this is the tower internals, either trays or packing. Therefore, it is common to encounter towers that are operating at the limit imposed by flooding.

The configuration in Figure 6.7 consists of a differential pressure-to-steam-flow cascade. The set point for the differential pressure controller (DPC) should be slightly below the differential pressure for the tower flooding. In

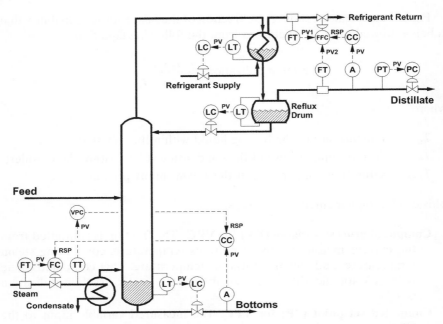

Figure 6.6. Valve position controller for condensing steam temperature.

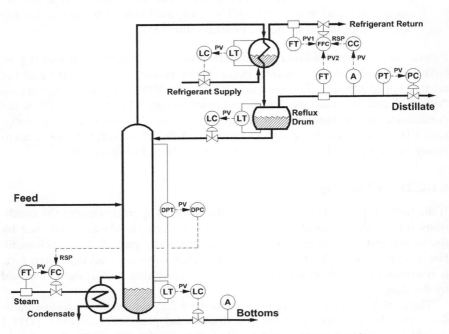

Figure 6.7. Differential-pressure-to-steam-flow cascade.

Figure 6.7, the differential pressure across both separation sections is measured. But when operating close to the flooding limit, the differential pressure is preferably measured across each separation section. The separation section where flooding will first occur is usually known. But occasionally, a selector configuration must be incorporated in order to take both separation sections into consideration.

The differential pressure measurement is the measured variable for a tower DPC that manipulates the steam to the reboiler. The tower DPC adjusts the set point for the reboiler steam flow controller. This configuration is frequently installed on towers and generally gives very good control of tower differential pressure.

The configuration in Figure 6.8 retains the bottoms composition controller via two levels of cascading:

- Tower differential pressure to bottoms composition.
- Bottoms composition to steam flow.

Differential-pressure-to-composition cascades are very unusual. For a cascade to function properly, the inner loop (the bottoms composition loop in the tower differential-pressure-to-bottoms-composition cascade) must be faster than the outer loop (the column differential pressure loop). The desire is for the inner loop to be faster than the outer loop by a factor of 5. For the

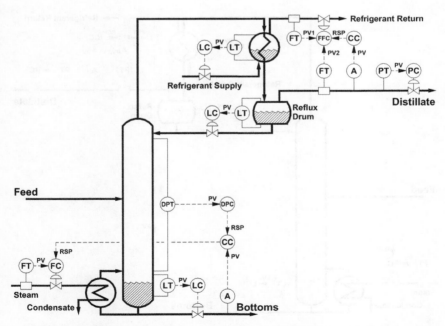

Figure 6.8. Differential-pressure-to-bottoms-composition-to-steam-flow cascade.

configuration in Figure 6.8, the separation of dynamics is in the wrong direction. The column differential pressure loop will certainly be faster than the bottoms composition loop.

For cascade, this is a disaster. Tuning problems will be encountered when tuning the column DPC. The controller will have to be tuned to respond very slowly, basically achieving the five-to-one separation in the required direction through a column DPC that responds very slowly. Since tower flooding is involved, tuning the column DPC in this manner is unacceptable.

6.4.5. Refrigerant Allocation

The refrigerant utilization is determined largely by the boilup. The boilup determines the overhead vapor flow. The distillate composition controller specifies the L/D ratio, which determines the fraction of the overhead vapor that must be condensed. This determines the heat transfer rate in the condenser, which determines the refrigerant flow.

And the response is relatively fast. Changing the heat to the reboiler quickly changes the vapor flow throughout the tower. To condense a constant fraction of the overhead vapor, the condenser must respond quickly.

The objective of the configuration in Figure 6.9 is to maintain the refrigerant flow at the amount allocated to the tower. The refrigerant flow is the measured variable for the refrigerant flow controller. The output of the

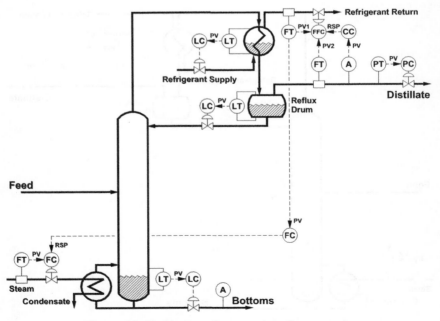

Figure 6.9. Refrigerant-flow-to-steam-flow cascade.

refrigerant flow controller is the set point to the steam flow controller. The resulting configuration is a flow-to-flow cascade.

The separation of dynamics required for cascade will be present:

- The steam flow controller exhibits the dynamics typical of flow loops.
- The refrigerant flow controller will respond far slower than the typical flow controller.

Responding to short-term variations in refrigerant flow is unnecessary. Instead of the instantaneous refrigerant flow, the measured variable for the refrigerant flow controller should be a period (perhaps 5 minutes) moving average of the refrigerant flow. A 1-second execution interval for the refrigerant flow controller is also unnecessary; executing on a 1-minute or 5-minute interval is adequate.

The configuration in Figure 6.10 is similar except that the refrigerant flow controller manipulates the set point of the bottoms composition controller. The result is a flow-to-composition-to-flow cascade, which will likely raise some eyebrows regarding the flow-to-composition layer of the cascade. Such configurations are extremely unusual, but then the refrigerant flow controller is also unusual in that it responds very slowly. But even so, it is unlikely that its response could be slowed to the point that it is five times slower that the bottoms composition controller.

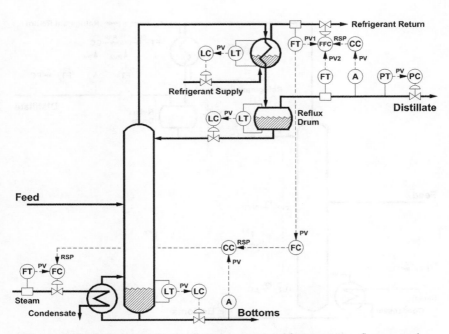

Figure 6.10. Refrigerant-flow-to-bottoms-composition-to-steam-flow cascade.

6.5. CONTROL CONFIGURATIONS FOR MULTIPLE CONSTRAINTS

The objective is to produce as much distillate product as possible subject to the following three constraints:

- Heat transfer in the reboiler.
- Tower flooding.
- Refrigerant allocation to the condenser.

The previous section presented configurations for each individual constraint; this section proposes configurations where all three must be incorporated. As used in these formulations, "constraint" is the actual constraint adjusted to provide the required operating margin.

6.5.1. Selector

Control personnel prefer to approach such problems with a signal selector as illustrated in Figure 6.11. To obtain the signal for the reboiler steam valve position, the low signal selector chooses the lesser of the following:

Output of the refrigerant flow controller. This controller is responsible for the constraint on refrigerant allocation.

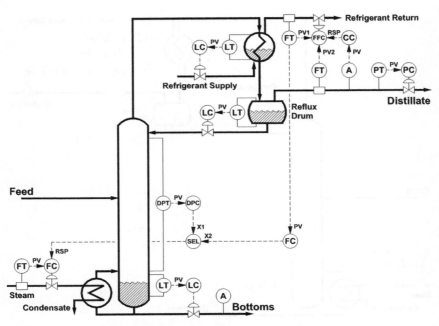

Figure 6.11. Selector for multiple constraints.

Output of the column DPC. This controller is responsible for the constraint on tower flooding.

Usually, there is an input for each constraint. But in this case, the third constraint (heat transfer limit in reboiler) corresponds to the reboiler steam valve fully open. A third input to the selector could be configured with a constant value of 100%, but it would serve no useful purpose. The limits on the final control element are automatically imposed on the output of the selector, so no input is required for this constraint.

On piping and instrumentation (P&I) diagrams, selectors are deceptively simple. To prevent windup in the controllers that provide the inputs to the selector, every controller whose input is not selected must be tracking the output of the selector. There are various techniques for doing this (external reset, integral tracking, inhibit increase/decrease), and all digital systems provide at least one of these. However, they must be configured properly in order to obtain a smooth transition from one controller to another.

6.5.2. Emulation of a Good Operator

A conscientious process operator can provide constraint control. The logic is actually quite simple:

DO on the hour:
  Check each constraint.
  IF any constraint is violated **THEN**
    Increase composition set point for light key in bottoms.
  ELSE IF at any constraint **THEN**
    Don't mess with it.
  ELSE IF below all constraints **THEN**
    Decrease composition set point for light key in bottoms.
  ENDIF
ENDDO

Procedures of this type are in the domain of expert systems. A set of rules can be formulated for the constraint control logic.

One of the deficiencies of the selector approach is that all constraints are treated equally. This is not quite the case. Violation of the tower flooding constraint has more serious consequences than violating the refrigerant allocation constraint. The refrigerant allocation limit can be exceeded by a considerable amount for a short duration without any consequences at all; it is the hourly average that is important. This is not the case for the tower flooding constraint, which must not be exceeded at any time. There are several possible enhancements to the above logic, such as checking the tower flooding constraint more frequently and increasing the bottoms composition set point by a greater amount if the flooding constraint is approached.

While expert systems have potential applications in constraint control, to date few applications have been reported. Initially, system issues impeded interfacing expert systems with digital process controls. These have been largely resolved, but with apparently little impact on applying expert systems for constraint control.

6.5.3. Truth Table

This implementation proceeds as follows:

- For each constraint, specify a tolerance to be "at" the constraint. This gives three possibilities: above the constraint, at the constraint, or below the constraint.
- Enumerate all possible combinations. For three constraints, there are 27 combinations. For n constraints, there are $3n$ combinations.
- Determine what action to take for each combination. For the tower, the possible actions are to increase the bottoms composition set point, to do nothing, or to decrease the bottoms composition set point.

The truth table in Table 6.1 enumerates all of the possibilities.

The only combination for which the bottoms composition set point can be decreased (less lights in the bottoms) is when the tower is below all constraints. For seven combinations, the action is to do nothing—no constraint is violated, but the tower is at one or more constraints. For all others, the set point must be increased. This implementation can also be "fine-tuned." For example, the set point might be increased by a greater amount if the tower flooding constraint is violated.

6.5.4. Model-Based Approach

To formulate the model-based approach (2), the first step is to express each constraint in terms of some common variable. For the tower, the common variable is the overhead vapor flow V_C. This flow is not directly measured, but it can be computed as the sum of the distillate flow D and the reflux flow L.

For each constraint, the limiting overhead vapor flow is computed from the current overhead vapor flow and the current conditions within the tower. The relationships are as follows:

Reboiler. Under heat transfer limited conditions, the current ΔT is the condensing steam temperature T_S less the bottoms temperature T_B. The maximum ΔT is the steam supply saturation temperature $T_{S,SAT}$ less T_B.

TABLE 6.1. Truth Table for Constraint Control

Column Flooding Constraint	Refrigerant Allocation Constraint	Reboiler Heating Constraint	Set Point for Impurities in Bottoms
Below	Below	Below	Decrease
At	Below	Below	No change
Above	Below	Below	Increase
Below	At	Below	No change
At	At	Below	No change
Above	At	Below	Increase
Below	Above	Below	Increase
At	Above	Below	Increase
Above	Above	Below	Increase
Below	Below	At	No change
At	Below	At	No change
Above	Below	At	Increase
Below	At	At	No change
At	At	At	No change
Above	At	At	Increase
Below	Above	At	Increase
At	Above	At	Increase
Above	Above	At	Increase
Below	Below	Above	Increase
At	Below	Above	Increase
Above	Below	Above	Increase
Below	At	Above	Increase
At	At	Above	Increase
Above	At	Above	Increase
Below	Above	Above	Increase
At	Above	Above	Increase
Above	Above	Above	Increase

The ratio of the limiting overhead vapor rate to the current overhead vapor is the ratio of the maximum heat transfer rate to the current heat transfer rate, which is the ratio of these ΔT's:

$$\frac{V_{C,REB}}{V_C} = \frac{Q_{MAX}}{Q} = \frac{T_{S,SAT} - T_B}{T_S - T_B},$$

where

$V_{C,REB}$ = limiting overhead vapor rate imposed by reboiler heat transfer;

V_C = current overhead vapor rate;

Q_{MAX} = maximum heat transfer rate in reboiler;

Q = current heat transfer rate in reboiler.

Tower flooding. The differential pressure ΔP is proportional to the square of the vapor flow. The maximum vapor flow corresponds to the pressure drop ΔP_{MAX} for flooding, adjusted to provide the operating margin:

$$\frac{V_{C,TWR}}{V_C} = \left[\frac{\Delta P_{MAX}}{\Delta P} \right]^{1/2},$$

where

$V_{C,TWR}$ = limiting overhead vapor rate imposed by flooding;

V_C = current overhead vapor rate;

ΔP_{MAX} = pressure drop for onset of flooding;

ΔP = current column pressure drop.

Condenser. The refrigerant flow R depends on the overhead vapor flow (a specified fraction must be condensed). The maximum vapor flow is determined by the refrigerant allocation R_{MAX}:

$$\frac{V_{C,CND}}{V_C} = \frac{R_{MAX}}{R},$$

where

$V_{C,CND}$ = limiting overhead vapor rate imposed by refrigerant allocation;

V_C = current overhead vapor rate;

R_{MAX} = refrigerant flow from refrigerant allocation;

R = current refrigerant flow.

Baxley [2] recommended an incremental implementation utilizing the following steps:

- Determine the limiting overhead vapor flow from the values computed using the above relationships:

$$V_{C,MAX} = \min\{V_{C,REB}, V_{C,TWR}, V_{C,CND}\}.$$

- Compute the change ΔV_C from the current overhead vapor flow:

$$\Delta V_C = V_{C,MAX} - V_C.$$

- Impose a maximum limit $\Delta V_{C,MAX}$ on the change:

$$\Delta V_C = \min\{\Delta V_C, \Delta V_{C,MAX}\},$$
$$\Delta V_C = \max\{\Delta V_C, -\Delta V_{C,MAX}\}.$$

- Target for overhead vapor flow is current vapor flow plus the change:

$$V_{C,SP} = V_C + \Delta V_C.$$

- Make adjustments on the tower to attain the target for the overhead vapor.

Baxley [2] retained the bottoms composition controller. To do so requires a mechanism to adjust the set point to the bottoms composition controller so as to attain the target for the overhead vapor flow. Baxley proposed using a proportional–integral–derivative (PID) controller whose PV is the computed value for the overhead vapor flow V_C and whose output is the set point to the bottoms composition controller. In effect, the result is a flow-to-composition cascade. These are certainly unusual. The flow loop would have to respond very slowly, but optimization is performed on a long time frame and often small increments are imposed on any change.

Another option is to eliminate the bottoms composition controller. There are two possibilities:

- Translate change in the overhead vapor ΔV_C to a change in the steam flow set point and add to the current value of the steam flow S to give a new value for the steam flow set point S_{SP}:

$$S_{SP} = S + k_S \, \Delta V_C.$$

The coefficient k_S is the unit of steam per unit of overhead vapor. A value for k_S can be obtained from the stage-by-stage separation model.
- Use a PID controller whose PV is the computed value for the overhead vapor flow V_C and whose output is the set point to the steam flow controller, the result being a flow-to-flow cascade. A change in the steam flow translates to a change in the overhead vapor flow V_C rather rapidly, but the dynamics will be slower than those of the steam flow controller.

Incremental calculations are relatively tolerant of errors in equations and coefficients. Two factors contribute to this:

- The calculations always reference the current conditions within the tower.
- All changes are restricted to small changes.

Consequently, errors are tolerated in both the equations and the coefficients (such as k_S) in the equations. Even the square root could be omitted from the tower flooding constraint without introducing significant error.

6.5.5. Model Predictive Control

Optimizing a distillation column involves two issues that model predictive control handles quite well:

- Interaction between process variables, notably the product compositions.
- Constraints, especially constraints on dependent variables such as the tower pressure drop.

When only one constraint is involved, the configurations in the previous section are usually straightforward to implement and commission. But as the number of constraints increases, the configurations become more complex. Commissioning and subsequent troubleshooting also become more challenging. Model predictive control is definitely worth considering for the application.

More discussion of model predictive control is provided toward the end of the next chapter.

REFERENCES

1   Smith, D. E., W. S. Stewart, and D. E. Griffin, Distill with composition control, *Hydrocarbon Processing*, 57(2), February 1978, 99–107.
2   Baxley, R. A., Local optimizing control for distillation, *Instrumentation Technology*, 16, 1969, 75–80.

7

DOUBLE-END COMPOSITION CONTROL

Single-end composition control is rather forgiving; double-end composition control is not. The selection of the control configuration must take into account the degree of interaction between the two composition loops.

The customary approach is to first decide how the two compositions will be controlled. There will always be some interaction between the distillate composition and the bottoms composition loops. However, some configurations will exhibit more interaction than others. A technique known as the relative gain can assess the degree of interaction for a proposed control configuration. The assessment of interaction can be based on either the product compositions or the control stage temperatures. The values for the relative gain will be different, but the final conclusions are generally the same.

7.1. DEFINING THE PROBLEM

The column in Figure 7.1 will be used as the basis for examining double-end composition control. The key aspects are as follows:

- The tower is a two-product tower.
- Both product streams are liquid (total condenser).

Distillation Control: An Engineering Perspective, First Edition. Cecil L. Smith.
© 2012 John Wiley & Sons, Inc. Published 2012 by John Wiley & Sons, Inc.

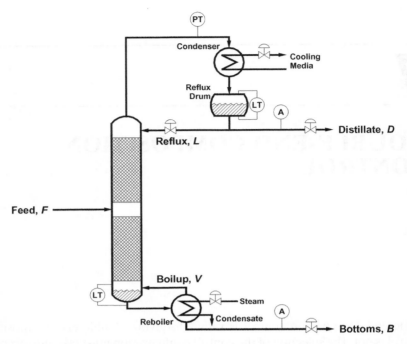

Figure 7.1. Controlled and manipulated variables for a two-product column.

- The condenser is not a flooded condenser, so the reflux drum level must be measured and controlled.
- Water is the cooling media for the condenser.
- Steam is the heating media for the reboiler.

7.1.1. Controlled and Manipulated Variables

Five variables must be controlled:

- Bottoms level.
- Reflux drum level.
- Column pressure.
- Distillate composition.
- Bottoms composition.

Both compositions are expressed as impurities in the product stream.

All measurements and all control valves are indicated on the tower in Figure 7.1. The manipulated variables in a process context are as presented in Table 1.3.

7.1.2. Multivariable Control Problem

There are five controlled variables and five manipulated variables. In multivariable control terminology, this is said to be a 5×5 "square" system.

In the single-loop approach to column control, a proportional–integral–derivative (PID) controller is configured for each of the controlled variables. The output of the controller must be to one of the control valves, but which one? The term "pairing" refers to the process of selecting the manipulated variable to be used for each controlled variable.

In columns with liquid product streams, the column pressure is almost always controlled by manipulating the heat transfer rate in the condenser. This will be assumed for the tower in Figure 7.1. This leaves four controlled variables and four manipulated variables.

7.1.3. Focus on Composition

The following observations are from a process operations perspective:

- The critical loops are the composition loops.
- The composition loops are the slowest of all.

The pressure control loop must be fast, especially if control stage temperatures are used in lieu of composition. In most cases, the level loops are fast relative to the composition loops. In effect, there is a reasonable dynamic separation between the composition loops and the other loops. This permits the pairing of loops to be approached as follows:

- Determine the appropriate manipulated variable for each of the composition loops.
- Use the remaining manipulated variables to control the two levels (assuming column pressure will be controlled by manipulating the heat transfer rate in the condenser).

7.2. OPTIONS FOR COMPOSITION CONTROL

The traditional approach to pairing controlled variables and manipulated variables can be summarized as "control each variable with the nearest valve that has a significant effect on that variable." For distillation columns, this statement translates to "control each variable with a manipulated variable on the same end of the tower." The options are as follows:

Distillate composition. The choices are reflux flow L or distillate flow D.
Bottoms composition. The choices are boilup V or bottoms flow B.

The total material balance $F = D + B$ must close, which imposes one restriction on the choice of the manipulated variables for the composition loops: either D or B may be a manipulated variable, but not both simultaneously. This leaves three choices for the manipulated variables: L, V, and D (or B). This gives three possible control configurations for the compositions.

7.2.1. Relative Gain

The appropriate control configuration depends on a number of factors, including

- the purity of the products,
- the external reflux ratio (L/D),
- the composition of the feed relative to the composition of the products.

These factors determine the success of copying a control configuration from one column to another. The copied configuration should work provided all of these factors are similar, but not if these factors are dissimilar.

How does one analyze such effects? By determining the degree of interaction through a technique known as the relative gain in combination with a stage-by-stage separation model of the column. Each of the two composition loops will be manipulating one of the final control elements. Therefore, this is a 2×2 multivariable system.

Although the basic principles are the same, the application of the relative gain to distillation columns is a little different:

- There are two controlled variables and two manipulated variables. But more than two manipulated variables are available. The result is multiple potential control configurations, each of which is a 2×2 multivariable system. The user begins by proposing a control configuration, and then using the relative gain to determine if that control configuration would function properly.
- In the usual application of the relative gain to a 2×2 multivariable system, reversing the pairing of the controlled and manipulated variables is an option. But for distillation, dynamic considerations make this option unacceptable.

The relative gain can only assess the interaction in a proposed configuration; it is of no help in proposing a control configuration for evaluation.

The objective is to find a control configuration with a low degree of interaction. Normally, one starts with the three simple configurations (to be presented shortly) that involve manipulating the reflux L, the boilup V, and either the distillate D or the bottoms B. If none of these are acceptable, then one proposes manipulating ratios such as the external reflux ratio L/D.

7.2.2. Notation for Control Configurations

Since various configurations will be proposed, the following convention for designating the configurations will be used:

$$X, Y \text{ configuration}$$

where

X = manipulated variable for controlling the distillate composition;

Y = manipulated variable for controlling the bottoms composition.

For example, with the L,B configuration the distillate composition y_H is controlled by manipulating the reflux flow L and the bottoms composition x_L is controlled by manipulating the bottoms flow B.

7.2.3. Distillate with L and Bottoms with V

The L,V configuration is illustrated in Figure 7.2. This is probably the most common configuration that is installed in production facilities. The 2×2 configuration is as follows:

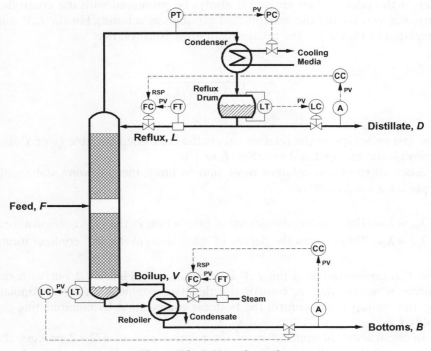

Figure 7.2. L,V configuration.

Controlled Variable	Manipulated Variable
Distillate composition y_H	Reflux flow L
Bottoms composition x_L	Boilup V

In many cases, this configuration either performs very poorly or not at all (one of the composition controllers remains on manual because of tuning difficulties). This configuration can be examined from the perspective of the material and energy balances:

Energy balance. Both L and V are related to energy. Their ratio (L/V) is the internal reflux ratio, which determines the separation provided by the tower. Both manipulated variables affect the separation, which leads to interaction.

Material balance. The two product flows are manipulated variables for level controllers. Each product flow is determined by the difference between the vapor flow and the liquid flow that are the manipulated variables for the two composition controllers.

7.2.4. Relative Gain Array

Herein the relative gain array will always be presented with the controlled variables as rows and the manipulated variables as columns. For the L,V configuration in Figure 7.2, the relative gain array is as follows:

	L	V
y_H	λ_{yL}	λ_{yV}
x_L	λ_{xL}	λ_{xV}

The first subscript on the relative gain is the controlled variable (y or x); the second is the manipulated variable (L or V).

Since all rows and columns must sum to unity, the following statements apply for a 2×2 process:

$\lambda_{yL} = \lambda_{xV}$. These assess the degree of interaction in the L,V configuration.
$\lambda_{xL} = \lambda_{yV}$. These assess the degree of interaction in the V,L configuration.

The V,L configuration is the L,V configuration in Figure 7.2 but with the pairing reversed, that is, control the distillate composition by manipulating the boilup V and control the bottoms composition by manipulating the reflux L.

In the traditional application of the relative gain to a 2×2 process, the pairing with the smallest relative gain is recommended. However, the relative gain only assesses the steady-state aspects of interaction. Unfortunately, configurations with a low degree of steady-state interaction might not be practical because of considerations relating to process dynamics. This is the case for the V,L configuration—it would not be considered even if the degree of interaction were zero.

This is one aspect where the application of the relative gain to distillation is different. Only the control configuration as proposed is viable. Reversing the pairing is not an option.

7.2.5. Distillate with D and Bottoms with V

The D,V configuration is illustrated in Figure 7.3. The 2×2 configuration is as follows:

Controlled Variable	Manipulated Variable
Distillate composition y_H	Distillate flow D
Bottoms composition x_L	Boilup V

This configuration can be examined from the perspective of the material and energy balances:

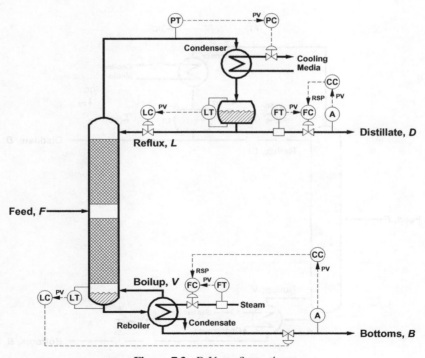

Figure 7.3. D,V configuration.

Energy balance. The bottoms composition controller works through the energy balance and separation. The manipulated variable is the boilup V; the reflux L is the difference between the overhead vapor flow and the distillate flow D.

Material balance. The distillate composition controller works through the material balance. The manipulated variable is the distillate flow D; the bottoms flow B is the difference between the feed flow F and the distillate flow D.

The relative gain array for the D,V configuration is as follows:

	D	V
y_H	λ_{yD}	λ_{yV}
x_L	λ_{xD}	λ_{xV}

7.2.6. Distillate with L and Bottoms with B

The L,B configuration is illustrated in Figure 7.4. The 2×2 configuration is as follows:

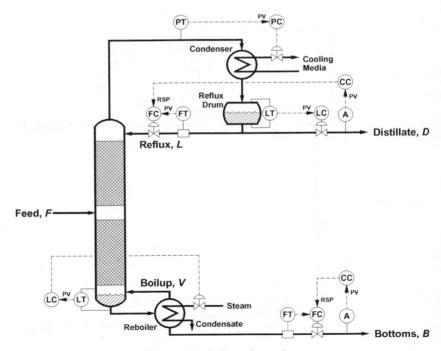

Figure 7.4. L,B configuration.

Controlled Variable	Manipulated Variable
Distillate composition y_H	Reflux flow L
Bottoms composition x_L	Bottoms flow B

This configuration can be examined from the perspective of the material and energy balances:

Energy balance. The distillate composition controller works through the energy balance and separation. The manipulated variable is the reflux flow L; the boilup V is the difference between the liquid flow from the lower separation section and the bottoms flow B.

Material balance. The bottoms composition controller works through the material balance. The manipulated variable is the bottoms flow B; the distillate flow D is the difference between the feed flow F and the bottoms flow B.

The relative gain array for the L,B configuration is as follows:

$$
\begin{array}{c|cc}
 & L & B \\
\hline
y_H & \lambda_{yL} & \lambda_{yB} \\
x_L & \lambda_{xL} & \lambda_{xB}
\end{array}
$$

The notation for the subscripts for the relative gains can give the wrong impression. The relative gain λ_{yL} appears in the relative gain arrays for the L,V configuration and the L,B configuration (in fact, it will appear in any relative gain array for which the reflux flow L is the manipulated variable for the distillate composition). However, the value of λ_{yL} in the L,V configuration is not the same as its value in the L,B configuration.

7.2.7. Ratio of Two Flows

In terms of the three basic manipulated variables L, V, and D (or B), only the three control configurations presented previously can be considered. However, configurations can be proposed for the manipulated variables being various ratios of these variables.

One such ratio is the external reflux ratio L/D. Consider the following 2×2 configuration:

Controlled Variable	Manipulated Variable
Distillate composition y_H	External reflux ratio L/D
Bottoms composition x_L	Boilup V

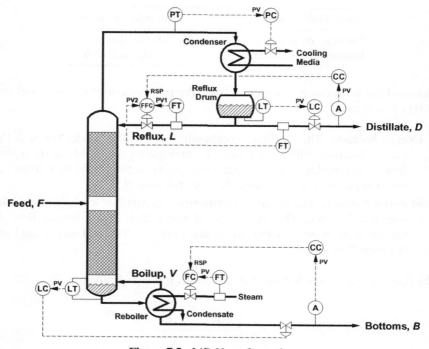

Figure 7.5. $L/D, V$ configuration.

The configuration in Figure 7.5 is for the ratio L/D. The flow-to-flow controller (FFC) ratios the reflux flow (the controlled flow) to the distillate flow (the wild flow). The distillate composition controller manipulates the set point for L/D.

The relative gain array for the $L/D, V$ configuration is as follows:

	L/D	V
y_H	$\lambda_{yL/D}$	λ_{yV}
x_L	$\lambda_{xL/D}$	λ_{xV}

When implementing the control configuration, the ratio D/L can be used instead of L/D. The degree of interaction is the same. As will be discussed later in this chapter, the choice of D/L or L/D is based on measurement noise and propagation of variance.

7.2.8. Composition versus Control Stage Temperature

The relative gain can assess the degree of interaction based on either the product composition or the respective control stage temperature. The values for the degree of interaction will be different.

However, the objective is to determine which control configuration is suitable for the column. Is it possible for a control configuration to the acceptable when based on temperatures but not acceptable when based on compositions, or vice versa? Probably, but this is very unusual.

As in Figures 7.2–7.5, the configurations will be presented only in terms of the product compositions. But in the subsequent examples for the depropanizer, the relative gains will be evaluated in terms of product compositions and in terms of control stage temperatures.

7.3. RELATIVE GAIN

Values for the relative gains will be computed using the stage-by-stage separation model. If undertaken during design, the base case will be the design basis. If undertaken for an operating column, the base case will be the normal operating conditions for the column.

Figure 1.14 provides the flows, compositions, and temperatures for the normal operating conditions for a depropanizer. As these values are from the solution of the stage-by-stage separation model, values for flows such as the boilup flow are available.

Computing the relative gains entails making changes in various flows, such as reflux, boilup, distillate, and bottoms. For linear systems, the results are independent of the magnitude and direction of the change. But distillation is nonlinear. The values selected for the change are a compromise between two issues:

- The larger the change, the greater the contribution of the column nonlinearities.
- The change must be sufficiently large that the resulting changes in the compositions are significantly larger than the "noise" associated with the convergence errors for the iterative solution of the stage-by-stage separation model.

7.3.1. Simple Concept

The relative gain is actually a simple concept, but the accompanying notation can give the impression of extreme complexity. Consider the distillate composition loop in Figure 7.2. The process gain or sensitivity for the distillate composition loop is the sensitivity of the distillate composition to changes in the reflux flow. But being one loop in a 2×2 multivariable system, this sensitivity can be evaluated in two different contexts:

Bottoms composition loop on manual. The boilup (the manipulated variable for the bottoms loop) is maintained at a fixed value.

Bottoms composition loop on automatic. The bottoms composition (the controlled variable for the bottoms loop) is maintained at a fixed value.

The relative gain is merely the ratio of these two sensitivities.

Why are these sensitivities important? Switching the bottoms composition controller between manual and automatic affects the process sensitivity of the distillation composition loop, which affects its performance. Assume the distillate composition controller is tuned first, which means that the bottoms composition controller is on manual during this tuning endeavor. Will the distillate composition loop perform properly when the bottoms composition loop is switched to automatic? Depends on the difference in the above two sensitivities. There are three cases:

- **Approximately the same (relative gain is approximately 1.0).** The distillate composition loop should function properly.
- **Very different (relative gain is much greater than 1.0 or much less than 1.0).** The distillate composition loop will not function properly. The only exception is if the dynamics of the two loops are very different, which is not the case for the distillate composition and bottoms composition loops.
- **Different sign (relative gain is negative).** If the distillate composition loop functions with the bottoms composition controller in manual, switching the bottoms composition loop to automatic will cause the distillate composition loop to be unstable.

This is the essence of the relative gain concept—basically a very simple concept.

7.3.2. Notation

This will be presented in the context of the L,V configuration in Figure 7.2. The relative gain array for this control configuration is

	L	V
y_H	λ_{yL}	λ_{yV}
x_L	λ_{xL}	λ_{xV}

Only one of the relative gains must be evaluated using the stage-by-stage separation model. Since all rows and columns must sum to unity, the remaining relative gains can be easily computed.

The relative gain λ_{yL} is defined as the ratio of two sensitivities:

$$\lambda_{yL} = \frac{K_{yL}}{K_{yL}},$$

where

K_{yL} = sensitivity of distillate composition y_H to the reflux flow L at constant boilup V (bottoms composition controller on manual);

K_{yL}' = sensitivity of distillate composition y_H to the reflux flow L at constant bottoms composition x_L (bottoms composition controller on auto).

These two sensitivities are formally expressed as partial derivatives that can be approximated by finite differences from which values can be computed from solutions of the stage-by-stage separation model:

$$K_{yL} = \left.\frac{\partial y_H}{\partial L}\right|_V \cong \left.\frac{\Delta y_H}{\Delta L}\right|_V,$$

$$K_{yL}' = \left.\frac{\partial y_H}{\partial L}\right|_{x_L} \cong \left.\frac{\Delta y_H}{\Delta L}\right|_{x_L}.$$

And to cap off the complex notation, these can be substituted into the expression for the relative gain to obtain the following:

$$\lambda_{yL} = \frac{K_{yL}}{K_{yL}'} = \frac{\left.\dfrac{\partial y_H}{\partial L}\right|_V}{\left.\dfrac{\partial y_H}{\partial L}\right|_{x_L}} \cong \frac{\left.\dfrac{\Delta y_H}{\Delta L}\right|_V}{\left.\dfrac{\Delta y_H}{\Delta L}\right|_{x_L}}.$$

Can anything this complex possibly be practical? The answer is definitely yes, but try convincing the skeptics. It is amazing that something so simple in concept can require such complex notation when expressed mathematically.

The relative gain certainly has limitations, the main ones being the following:

The relative gain only assesses the steady-state degree of interaction. As a result, the relative gain can recommend configurations that cannot function because of adverse dynamics. This arises in distillation, but they can be easily dismissed.

The relative gain is a linear systems concept. The impact on distillation will be discussed shortly. But be careful arbitrarily dismissing everything linear. The PID controller is linear; model predictive control (MPC) is also linear.

7.3.3. Control Configuration

The starting point is to choose a control configuration. For an operating column, the currently installed control configuration is usually chosen.

Otherwise, the chosen configuration is one that involves the basic manipulated variables, the options being as follows:

Configuration	L,V (Fig. 7.2)	D,V (Fig. 7.3)	L,B (Fig. 7.4)
Controlled variable C_1	y_H	y_H	y_H
Controlled variable C_2	x_L	x_L	x_L
Manipulated variable M_1	L	D	L
Manipulated variable M_2	V	V	B

For reasonably well-behaved towers like depropanizers, experienced people can often suggest the most appropriate configuration based on product purities, external reflux ratios, and so on. Especially in the chemical industry, separations often involve materials that deviate considerably from ideal. For such towers, basing the analysis on the results of the stage-by-stage separation model is the preferable approach.

The L,V configuration will be considered first, only because it is the most frequently installed.

7.3.4. Sensitivity of Distillate Composition y_H to Reflux L at Constant Boilup V

This requires two solutions of the stage-by-stage model:

	Base Case	Change Reflux	
V, mol/h	64.88	64.88	Constant V (bottoms CC on manual)
L, mol/h	57.00	57.10	$\Delta L = 0.10$ mol/h
y_H, mol%	0.0281	0.0263	Computed by stage-by-stage separation model

Figure 1.14 provides the values for the base case. The solution for the increase in the reflux must be computed for the specified values of L and V using the stage-by-stage separation model. Values for L and V can be specified directly to most stage-by-stage separation models. The only value of interest from the solution is the composition of the heavy key in the distillate.

The sensitivity K_{yL} is computed as follows:

$$K_{yL} = \frac{\partial y_H}{\partial L}\bigg|_V \cong \frac{\Delta y_H}{\Delta L}\bigg|_V = \frac{(0.0263 - 0.0281)\,\text{mol}\%}{(57.10 - 57.00)\,\text{mol/h}} = -0.018\,\text{mol}\%/(\text{mol/h})$$

7.3.5. Sensitivity of Distillate Composition y_H to Reflux L at Constant Bottoms Composition x_L

This requires two solutions of the stage-by-stage model:

	Base Case	Change Reflux	
x_L, mol%	0.7855	0.7855	Constant x_L (bottoms CC on automatic)
L, mol/h	57.00	57.10	$\Delta L = 0.10$ mol/h
y_H, mol%	0.0281	0.0274	Computed by stage-by-stage separation model

Figure 1.14 provides the values for the base case. The solution for the increase in the reflux must be computed for the specified values of L and x_L using the stage-by-stage separation model. Some stage-by-stage separation models permit the bottoms composition to be specified directly, but for some, the value of the boilup V must be varied until the computed value for the bottoms composition is the desired value. The only value of interest from the solution is the composition of the heavy key in the distillate.

The sensitivity K_{yL}' is computed as follows:

$$K_{yL}' = \left.\frac{\partial y_H}{\partial L}\right|_{x_L} \cong \left.\frac{\Delta y_H}{\Delta L}\right|_{x_L} = \frac{(0.0274 - 0.0281)\,\text{mol\%}}{(57.10 - 57.00)\,\text{mol/h}} = -0.007\,\text{mol\%/(mol/h)}.$$

When computing the values for the sensitivities K_{yL} and K_{yL}', the same values are usually used for the following:

- The base case.
- The increment ΔL.

However, different values can be used if desired.

7.3.6. Relative Gain λ_{yL}

The value for this relative gain is computed as follows:

$$\lambda_{yL} = \frac{K_{yL}}{K_{yL}'} = \frac{-0.018\,\text{mol\%/(mol/h)}}{-0.007\,\text{mol\%/(mol/h)}} = 2.6.$$

The sensitivity of distillate composition y_L to the reflux flow L changes by more than a factor of two when the bottoms composition controller is switched between manual and automatic. The sensitivity with the bottoms controller in manual is more than twice the sensitivity with the bottoms controller in automatic.

This suggests a substantial degree of interaction between the loops. The relative gain array for the L,V configuration is as follows:

	L	V			L	V
y_H	λ_{yL}	λ_{yV}	$=$	y_H	2.6	-1.6
x_L	λ_{xL}	λ_{xV}		x_L	-1.6	2.6

The desire is for the largest relative gain on each row and each column to be "near" 1.0. Unfortunately, "near" is not precisely defined. For a 2×2 process, values as low as 0.8 and as high as 1.2 are almost always satisfactory. Often values as low as 0.7 and as high as 1.4 are acceptable. However, 2.6 definitely does not qualify as "near."

7.3.7. Alternate Evaluation of a Relative Gain

The previous approach evaluated the relative gain λ_{yL} for controlling the distillate composition y_H by manipulating the reflux flow L. Selecting this relative gain to evaluate is quite arbitrary. In theory, one could evaluate any of the relative gains in the relative gain array and then compute the remainder from the fact that each row and each column must sum to unity. In practice, one would select one of the relative gains (λ_{yL} and λ_{xV}) on the diagonal as these correspond to the control loops in the L,V configuration in Figure 7.2.

Repeat the calculations for λ_{xV}, which is defined as the ratio of two sensitivities:

$$\lambda_{xV} = \frac{K_{xV}}{K_{xV}'},$$

where

K_{xV} = sensitivity of bottoms composition x_L to the boilup V at constant reflux flow L (distillate composition controller on manual);

K_{xV}' = sensitivity of distillate bottoms composition x_L to the boilup V at constant distillate composition y_H (distillate composition controller on auto).

These two sensitivities are expressed as follows:

$$K_{xV} = \left.\frac{\partial x_L}{\partial V}\right|_L \cong \left.\frac{\Delta x_L}{\Delta V}\right|_L,$$

$$K_{xV}' = \left.\frac{\partial x_L}{\partial V}\right|_{y_H} \cong \left.\frac{\Delta x_L}{\Delta V}\right|_{y_H}.$$

As for λ_{yL}, each sensitivity is computed from values obtained from solutions of the stage-by-stage model.

7.3.8. Sensitivity of Bottoms Composition x_L to Boilup V at Constant Reflux L

This requires two solutions of the stage-by-stage model:

	Base Case	Change Boilup	
L, mol/h	57.00	57.00	Constant L (distillate CC on manual)
V, mol/h	64.88	64.98	$\Delta V = 0.10$ mol/h
x_L, mol%	0.7855	0.7467	Computed by stage-by-stage separation model

Figure 1.14 provides the values for the base case. The solution for the increase in the boilup must be computed for the specified values of L and V using the stage-by-stage separation model. The only value of interest from the solution is the composition of the light key in the bottoms.

The sensitivity K_{xV} is computed as follows:

$$K_{xV} = \left.\frac{\partial x_L}{\partial V}\right|_L \cong \left.\frac{\Delta x_L}{\Delta V}\right|_L = \frac{(0.7467 - 0.7855)\,\text{mol\%}}{(64.98 - 64.88)\,\text{mol/h}} = -0.388\;\text{mol\%/(mol/h)}.$$

7.3.9. Sensitivity of Bottoms Composition x_L to Boilup V at Constant Distillate Composition y_H

This requires two solutions of the stage-by-stage model:

	Base Case	Change Boilup	
y_H, mol%	0.0281	0.0281	Constant y_H (distillate CC on automatic)
V, mol/h	64.88	64.98	$\Delta V = 0.10$ mol/h
x_L, mol%	0.7855	0.7705	Computed by stage-by-stage separation model

Figure 1.14 provides the values for the base case. The solution for the increase in the boilup must be computed for the specified values of V and y_H using the stage-by-stage separation model. The only value of interest from the solution is the composition of the light key in the bottoms.

The sensitivity K_{xV}' is computed as follows:

$$K_{xV}' = \left.\frac{\partial x_L}{\partial V}\right|_{y_H} \cong \left.\frac{\Delta x_L}{\Delta V}\right|_{y_H} = \frac{(0.7705 - 0.7855)\,\text{mol\%}}{(64.98 - 64.88)\,\text{mol/h}} = -0.150\;\text{mol\%/(mol/h)}.$$

7.3.10. Relative Gain λ_{xV}

The value for this relative gain is the ratio of these two sensitivities:

$$\lambda_{xV} = \frac{K_{xV}}{K_{xV}'} = \frac{-0.388 \text{ mol\%}/(\text{mol/h})}{-0.150 \text{ mol\%}/(\text{mol/h})} = 2.6.$$

The value of λ_{xV} is the same as determined previously. However, this is definitely not assured.

For linear systems, the results should be identical. But for a nonlinear process such as distillation, obtaining such agreement requires two actions:

1. Very small changes must be made in L and V. In the above examples, the change was 0.1 mol/h, which is approximately 0.2% of the respective values.
2. Very small tolerances must be specified for the convergence of the stage-by-stage separation model.

The precision to which the results of the stage-by-stage separation model are expressed are almost certainly beyond the accuracy of the model. However, the sensitivities are computed from changes, not from the actual values. This is similar to the accuracy and repeatability attributes of measurement devices—the repeatability is usually much better than the accuracy.

7.4. RELATIVE GAINS FROM OPEN LOOP SENSITIVITIES

For 2×2 processes, values for the relative gains can be computed as the ratio of the two sensitivities in the formal definition of the relative gain. For higher dimensional processes, this approach is impractical. For distillation, the two composition loops in a two-product tower constitute a 2×2 process. The addition of a sidestream to the tower results in three composition loops and a 3×3 process. Such towers require a different approach.

The alternate approach involves computing the process gain matrix $\mathbf{K}$ in the equation

$$\mathbf{c} = \mathbf{K} \, \mathbf{m},$$

where

$\mathbf{c}$ = vector of changes in the controlled variables;

$\mathbf{m}$ = vector of changes in the manipulated variables;

$\mathbf{K}$ = process gain matrix.

For the L, V configuration for a distillation column, this equation can be written as follows:

$$\begin{bmatrix} \Delta y_H \\ \Delta x_L \end{bmatrix} = \begin{bmatrix} K_{yL} & K_{yV} \\ K_{xL} & K_{xV} \end{bmatrix} \begin{bmatrix} \Delta L \\ \Delta V \end{bmatrix}$$

$\mathbf{c} = \begin{bmatrix} \Delta y_H \\ \Delta x_L \end{bmatrix}$ = vector of changes in the controlled variables;

$\mathbf{m} = \begin{bmatrix} \Delta L \\ \Delta V \end{bmatrix}$ = vector of changes in the manipulated variables;

$\mathbf{K} = \begin{bmatrix} K_{yL} & K_{yV} \\ K_{xL} & K_{xV} \end{bmatrix}$ = process gain matrix.

These equations provide a linear approximation for the column, which must be used with caution as distillation is quite nonlinear.

The relative gains are computed from the process gain matrix using the following equation:

$$\lambda_{ij} = K_{ij} \times (\mathbf{K}^{-1})^T{}_{ij}.$$

The computational procedure is as follows:

- Use the stage-by-stage separation model to compute the sensitivities. For each manipulated variable $j, j = 1, 2, \ldots n$, do the following:
  1. Make a small change from the base case.
  2. Compute the solution of the stage-by-stage separation model.
  3. Determine the change in each product composition.
  4. Compute the sensitivity $K_{ij}, i = 1, 2, \ldots n$.
- Compose the process gain matrix $\mathbf{K}$ from the individual process gains K_{ij}.
- Compute the matrix inverse $\mathbf{K}^{-1}$ of the process gain matrix $\mathbf{K}$. If this inverse does not exist, then there are inadequate degrees of freedom for the proposed control configuration.
- Transpose the inverse of the process gain matrix. This is $(\mathbf{K}^{-1})^T$.
- To obtain λ_{ij}, multiply element i,j of $(\mathbf{K}^{-1})^T$ by K_{ij}, This is not matrix multiplication; it is an element-by-element product.

Although mandatory for 3×3 and higher dimensional processes, this approach can be applied to a 2×2 multivariable process such as the depropanizer. Table 7.1 presents the values for three solutions:

- Base case. Values are provided by Figure 1.14.
- Increase L at constant V. Computed from stage-by-stage separation model.

TABLE 7.1. Solutions for L,V Configuration

		Change Reflux ($\Delta L = +0.10$)	Change Boilup ($\Delta V = +0.10$)
L, mol/h	57.00	57.10	57.00
V, mol/h	64.88	64.88	64.98
y_H, mol%	0.0281	0.0263	0.0297
x_L, mol%	0.7855	0.8170	0.7467
T_6, °C	50.44	50.25	50.62
T_{17}, °C	106.13	105.86	106.45

- Increase V at constant L. Computed from stage-by-stage separation model.

Table 7.1 provides values for the control stage temperatures as well as the product compositions.

7.4.1. Increase L at Constant V

A small change ΔL is made in the reflux flow L and values of y_H and x_L computed using the stage-by-stage separation model (the "Change Reflux" case in Table 7.1). From this solution, the two sensitivities to a change in the reflux flow L are evaluated as follows:

$$K_{yL} = \frac{\partial y_H}{\partial L}\bigg|_V \cong \frac{\Delta y_H}{\Delta L}\bigg|_V = \frac{(0.0263 - 0.0281)\,\text{mol\%}}{(57.10 - 57.00)\,\text{mol/h}} = -0.018\,\text{mol\%/(mol/h)},$$

$$K_{xL} = \frac{\partial x_L}{\partial L}\bigg|_V \cong \frac{\Delta x_L}{\Delta L}\bigg|_V = \frac{(0.8170 - 0.7855)\,\text{mol\%}}{(57.10 - 57.00)\,\text{mol/h}} = 0.315\,\text{mol\%/(mol/h)}.$$

7.4.2. Increase V at Constant L

A small change ΔV is made in the boilup V and values of y_H and x_L computed using the stage-by-stage separation model (the "Change Boilup" case in Table 7.1). From this solution, the two sensitivities to a change in the boilup V are evaluated as follows:

$$K_{yV} = \frac{\partial y_H}{\partial V}\bigg|_L \cong \frac{\Delta y_H}{\Delta V}\bigg|_L = \frac{(0.0297 - 0.0281)\,\text{mol\%}}{(64.98 - 64.88)\,\text{mol/h}} = 0.016\,\text{mol\%/(mol/h)},$$

$$K_{xV} = \frac{\partial x_L}{\partial V}\bigg|_L \cong \frac{\Delta x_L}{\Delta V}\bigg|_L = \frac{(0.7467 - 0.7855)\,\text{mol\%}}{(64.98 - 64.88)\,\text{mol/h}} = -0.388\,\text{mol\%/(mol/h)}.$$

7.4.3. Compute the Relative Gains

The process gain matrix is composed from the values for the four sensitivities computed above:

$$K = \begin{bmatrix} -0.018 & 0.016 \\ 0.315 & -0.388 \end{bmatrix}.$$

Computing $(K^{-1})^T$ and multiplying each element by the respective element of K gives the following relative gain array:

	L	V
y_H	3.5	-2.5
x_L	-2.5	3.5

This suggests an even larger degree of interaction between the loops (a relative gain of 3.5 as compared with 2.6). However, the conclusion is the same: the L,V configuration will exhibit substantial interaction between the loops.

7.4.4. Relative Gains for Temperatures

When evaluating the relative gains for a proposed control configuration, the relative gains can be based on composition measurements, temperature measurements, or a combination (e.g., distillate on composition, bottoms on temperature). The resulting numerical values for the relative gains are generally different. However, the conclusion of interest is the viability of the proposed control configuration. Rarely is the conclusion different.

Evaluating the relative gains based on temperatures entails computing the same solutions as for the compositions. Table 7.1 captures the values for both product compositions and control stage temperatures. These values give the following relative gain array:

	L	V
T_6	4.1	-3.1
T_{17}	-3.1	4.1

The value of 4.1 for the relative gain based on temperatures compares favorably with the value of 3.5 for the relative gain based on compositions. Both suggest substantial interaction between the loops in the L,V configuration.

7.4.5. Effect of Increment for Finite Difference Approximation

In all previous examples, the reflux and the boilup were increased by 0.1 mol/h from the base case to obtain the solutions for evaluating the sensitivities. This change is slightly over 0.2% of the value of L and V. This small increment gave small differences in both compositions and temperatures, which requires a tight convergence tolerance for the stage-by-stage separation model.

Distillation is a nonlinear process and one must pay careful attention to the increment size. For an increment size of 0.5 mol/h (approximately 1% of the value of L and V) is used, the values computed for the relative gains are very different. When computed from the open loop sensitivities, the relative gain arrays for compositions and temperatures are as follows:

	L	V		L	V
y_H	−3.2	4.2	T_6	−5.4	6.4
x_L	4.2	−3.2	T_{17}	6.4	−5.4

The relative gains on the diagonal are now negative, whereas formerly they were positive. The only way to be certain that the value for the increment is appropriate is to try smaller and smaller values until the values of the relative gains do not change significantly.

7.5. RELATIVE GAINS FOR OTHER CONFIGURATIONS

Why was the L,V configuration chosen as the starting point? Because it continues to be the one most frequently installed. In most applications, the L,V configuration will exhibit more interaction than the alternatives.

Given that the L,V configuration exhibits substantial interaction, which configuration should be analyzed next? By examining the values of the relative gains for the L,V configuration, is it possible to suggest which configuration to analyze next? Unfortunately, the answer is no.

All of the results in this section are based on relative gains calculated from the open-loop process sensitivities (as in the previous section). Using the approaches in Section 7.3 will give different values from the relative gains, but the conclusions will be the same.

Which configuration should be analyzed next? Usually, one exhausts the simple configurations before proceeding to configurations involving ratios such as the external reflux ration L/D. That means either the D,V configuration or the L,B configuration. The D,V configuration is arbitrarily selected.

The base case is always the same as for the L,V configuration. Relative gain arrays for both composition and temperature will be computed.

7.5.1. *D,V* Configuration

The following solutions are required:

1. Base case.
2. Solution for a change in D, but the same value of V as in the base case.
3. Solution for a change in V, but the same value of D as in the base case.

TABLE 7.2. Solutions for D,V Configuration

		Change Distillate ($\Delta D = +0.10$)	Change Boilup ($\Delta V = +0.10$)
D, mol/h	22.80	22.90	22.80
V, mol/h	64.88	64.88	64.98
y_H, mol%	0.0281	0.0387	0.0273
x_L, mol%	0.7855	0.6600	0.7853
T_6, °C	50.44	51.57	50.35
T_{17}, °C	106.13	107.28	106.11

The values for each solution are presented in Table 7.2. These values give the following relative gain arrays:

	D	V		D	V
y_H	0.02	0.98	T_6	−0.3	1.3
x_L	0.98	0.02	T_{17}	1.3	−0.3

The values are different, but the conclusion is the same: the D,V configuration is not a good idea. The relative gain arrays suggest reversing the pairing—control y_H or T_6 by manipulating V and control x_L or T_{17} by manipulating D. But this is only from the steady-state perspective; dynamically this makes no sense.

7.5.2. L,B Configuration

When the D,V configuration suggests that the loop pairing should be reversed, the L,B configuration usually (but not always) looks good. The following solutions are required:

1. Base case.
2. Solution for a change in L, but the same value of B as in the base case.
3. Solution for a change in B, but the same value of L as in the base case.

The values for each solution are presented in Table 7.3. These values give the following relative gain arrays:

	L	B		L	B
y_H	0.99	0.01	T_6	1.1	−0.1
x_L	0.01	0.99	T_{17}	−0.1	1.1

Both suggest that the L,B configuration should perform properly. The relative gain array for compositions suggests almost no interaction between the

TABLE 7.3. Solutions for L,B Configuration

		Change Reflux ($\Delta L = +0.10$)	Change Bottoms ($\Delta B = +0.10$)
L, mol/h	57.00	57.10	57.00
B, mol/h	77.20	77.20	77.30
y_H, mol%	0.0281	0.0274	0.0243
x_L, mol%	0.7855	0.7853	0.9127
T_6, °C	50.44	50.37	50.00
T_{17}, °C	106.13	106.11	105.13

loops. The relative gain array for temperatures suggests a small degree of interaction, but a value of 1.1 for the relative gains on the diagonal is definitely "close to one."

7.6. RATIOS FOR MANIPULATED VARIABLES

Engineers at least give lip service to the "keep it simple" principle. For column control configurations, that means use configurations that rely on the following manipulated variables:

D or B
L
V

The possible configurations are

- The L,V configuration (Fig. 7.2).
- The D,V configuration (Fig. 7.3).
- The L,B configuration (Fig. 7.4).

For most columns, one of these will be satisfactory.

But what if none of the three basic configurations is satisfactory? One can consider various ratios of the basic manipulated variables, such as the L/D configuration in Figure 7.5.

7.6.1. Ratios of D (or B), L, and V

A good starting point is to develop a list of the various ratios that could be used as manipulated variables for controlling the compositions. Enumerating all possible ratios of D, L, and V gives the following ratios:

L/D

V/D

V/L

In any of the ratios, D can be replaced by B. This probably makes sense for V/D, after which the list becomes the following:

L/D—the external reflux ratio;

V/B—the boilup ratio;

V/L—the internal reflux ratio in the upper separation section.

All of these ratios are energy terms and are related. The equations are simplest for a tower for which

- the feed is at its bubble point and
- equimolal overflow can be assumed.

With these assumptions, the vapor flow up the tower is constant (the boilup flow equals the overhead vapor flow). A material balance around the condenser/reflux drum gives the following equation:

$$V = L + D.$$

Dividing by L gives an equation that algebraically relates the internal reflux ratio V/L in the upper separation section and the external reflux ratio L/D:

$$\frac{V}{L} = 1 + \frac{D}{L}.$$

A material balance around the reboiler gives the following equation:

$$L + F = B + V.$$

Dividing by B gives the following equation that relates the internal reflux ratio $V/(L + F)$ in the lower separation section and the boilup ratio V/B:

$$\frac{L + F}{V} = \frac{B}{V} + 1.$$

Consequently, only one of these ratios may be used as a manipulated variable in a control configuration. For example, either of the following is possible:

- Control the distillate composition y_H by manipulating the external reflux ratio L/D.

- Control the bottoms composition x_L by manipulating the boilup ratio V/B.

However, a given control configuration may not contain both.

7.6.2. A Ratio or Its Reciprocal

The external reflux ratio is normally expressed as L/D; the boilup ratio is normally expressed as V/B; the internal reflux ratio is expressed as V/L. However, the reciprocal of any of these ratios may be used within the control configuration.

The degree of interaction between a bottoms composition loop and either of the following distillation composition loops is exactly the same:

- A distillate composition loop manipulating the L/D ratio.
- A distillate composition loop manipulating the D/L ratio.

Consequently, the choice of L/D or D/L cannot be based on interaction. Instead, it is based on the propagation of the variance associated with the noise on a flow measurement.

Normally, the preference is to control using a ratio whose value is less than 1.0. Using L/D for example, the control configuration would be as follows:

- Provide a measurement for the distillate flow D.
- Provide a measurement and a controller for the reflux flow L.
- The output of the distillate composition controller is the desired value for the L/D ratio. As notation, L/D is the actual reflux-to-distillate flow ratio; $(L/D)_{SP}$ is the desired value for the L/D ratio.
- Multiply the measured value of the distillate flow D by the desired reflux-to-distillate flow ratio $(L/D)_{SP}$ to obtain the set point L_{SP} for the reflux flow controller.

As for any flow measurement, some noise accompanies the measured value of the distillate flow D. The preference is for the $(L/D)_{SP}$ to be less than 1.0 so that this noise is attenuated. If the $(L/D)_{SP}$ is greater than 1.0, the noise in the measured value of D is amplified.

7.6.3. Manipulated Variables for Compositions

Including the ratios, the list of potential manipulated variables is now as follows:

D or B

L

V

L/D or V/B or V/L

The logical options for controlling the distillate composition y_H are the following:

D

L

L/D or V/L

The logical options for controlling the bottoms composition x_L are the following:

B

V

V/B or V/L

There are some combinations that cannot be used:

- Control the distillate composition y_H using D and the bottoms composition x_L using B.
- Control the distillate composition y_H using L/D or V/L and the bottoms composition x_L using V/B or V/L.

7.6.4. Relative Gain for the $L/D,B$ Configuration

The relative gain for this configuration is computed in the same manner as for previous examples. For the depropanizer whose data are provided in Figure 1.14, the external reflux ratio L/D is 2.5. Since this ratio is greater than 1.0, the ratio D/L is preferred in the implementation of the control configuration. But as the degree of interaction is the same, the relative gain analysis can be based on L/D. The stage-by-stage separation model is free of measurement noise, so either ratio can be used.

The data for the $L/D,B$ configuration is obtained starting with values from the base case in Figure 1.14. The following solutions are required:

1. Base case.
3. Solution for a change in L/D, but the same value of B as in the base case.
2. Solution for a change in B, but the same value of L/D as in the base case.

The values for each solution are presented in Table 7.4. These values give the following relative gain arrays:

TABLE 7.4. Solutions for $L/D,B$ Configuration

		Change Reflux Ratio $\Delta(L/D) = +0.10$	Change Bottoms $\Delta B = +0.10$
L/D	2.50	2.51	2.50
B, mol/h	77.20	77.20	77.30
y_H, mol%	0.0281	0.0266	0.0257
x_L, mol%	0.7855	0.7850	0.9131
T_6, °C	50.44	50.28	50.16
T_{17}, °C	106.13	106.09	105.18

	L/D	B		L/D	B
y_H	0.99	0.01	T_6	1.1	−0.1
x_L	0.01	0.99	T_{17}	−0.1	1.1

The results are essentially identical to the results for the L,B configuration. There is no incentive to using the $L/D,B$ configuration instead of the L,B configuration.

7.7. EFFECT OF OPERATING OBJECTIVES

For the solution in Figure 1.14, the operating objectives are as follows:

- Maximize the amount of ethane (1.754 mol%) in the propane. To do this, the amount of butane in the propane is very small (0.0281 mol%).
- Maximize the amount of propane (0.7855 mol%) in the butane.

This makes sense when ethane is less valuable than propane, which is less valuable than butane.

What if the product values are reversed, that is, ethane is more valuable than propane, which is less valuable than butane? This is reflected in the solution in Figure 7.6. First, note that the feed now contains 0.1 mol% ethane instead of 0.4 mol% as in Figure 1.14. The operating objective becomes the following:

- Maximize the amount of butane in the propane. To do this, the amount of ethane in the feed to the depropanizer is much smaller.
- Minimize the amount of propane in the butane.

7.7.1. Relative Gains for the L,B Configuration

This configuration will be examined initially because it was the preferable configuration for the previous operating objectives. For the revised operating

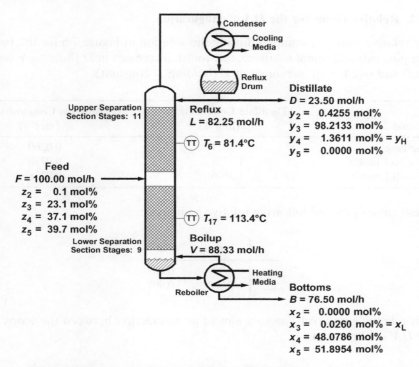

Figure 7.6. Depropanizer model for maximizing butane in propane.

objectives, the relative gains are computed using the values from Figure 7.6 as the base case plus two additional solutions, one for an increment in L (holding B constant) and one for an increment in B (holding L constant):

Solution	Distillate Composition y_H (mol%)	Bottoms Composition x_L (mol%)
Base case	1.3611	0.0260
$\Delta L = +0.1$ mol/h	1.3606	0.0258
$\Delta B = +0.1$ mol/h	0.9428	0.0269

These values give the following relative gain array:

	L	B
y_H	0.01	0.99
x_L	0.99	0.01

This relative gain array suggests that the L,B configuration is inappropriate.

7.7.2. Relative Gains for the D,V Configuration

The relative gains are computed using the solution in Figure 7.6 for the base case plus two additional solutions, one for an increment in D (holding V constant) and one for an increment in V (holding D constant):

Solution	Distillate Composition y_H (mol%)	Bottoms Composition x_L (mol%)
Base case	1.3611	0.0260
$\Delta D = +0.1$ mol/h	1.7778	0.0256
$\Delta V = +0.1$ mol/h	1.3606	0.0258

These values give the following relative gain array:

	D	V
y_H	0.997	0.003
x_L	0.003	0.997

This relative gain array suggests almost no interaction between the loops in the D,V configuration.

7.7.3. Relative Gains for the L,V Configuration

The relative gains are computed using the solution in Figure 7.6 for the base case, two additional solutions are required, one for an increment in L (holding V constant) and one for an increment in V (holding L constant):

Solution	Distillate Composition y_H (mol%)	Bottoms Composition x_L (mol%)
Base case	1.3611	0.0260
$\Delta D = +0.1$ mol/h	1.2312	0.0261
$\Delta V = +0.1$ mol/h	1.5180	0.0256

These values give the following relative gain array:

	D	V
y_H	1.9	−0.9
x_L	−0.9	1.9

This relative gain array suggests that the interaction between the loops in the L,V configuration is too high for this configuration to perform satisfactorily.

7.7.4. Observations

When a significant change in the operating conditions is made, retuning the controllers is generally anticipated. However, this example suggests that the change may have to go beyond controller tuning. Modifications may be required to the control configuration itself.

For this example, the changes in operating conditions are price driven, that is, by the prices of ethane, propane, and butane. These prices could change at any time. But more likely, they are the result of

- the long-term trends in the prices,
- short-term effects such as the season of the year, and
- technological shifts in the use of the product.

Reacting to such changes is the objective of plant optimization efforts. Normally, it is assumed that the response can be implemented via adjusting the targets of the existing control systems. This example suggests that this is not always the case.

7.8. MPC

To most people, controlling distillate composition with reflux and bottoms composition with boilup (the L,V configuration) seems to be the most appropriate approach. But in practice, controlling one product composition with an energy term (L or V) and the other with a product draw (D or B) is usually preferred because of a lower degree of interaction between the control loops. By providing a quantitative value on the degree of interaction between the loops, the relative gain hopefully converts what is often a largely qualitative discussion to one based on quantitative data.

Getting changes to a piping and instrumentation (P&I) diagram implemented is rarely easy. Discussions of the changes are definitely appropriate, and the reasons for making any change should be documented. However, the discussions can go on and on with two possible outcomes:

We decide not to decide. Decisions can be avoided by simply continuing the discussions.

The pocket veto. A decision is made to change the configuration, but it is never implemented.

This creates interest in technologies that circumvent these discussions.

One possibility is MPC. Distillation has two and possibly three attributes that make MPC a potential candidate for providing the control:

Distillation is multivariable. The compositions present a 2×2 multivariable process. Including the levels and pressure present a 5×5 multivariable process.

Distillation is an interacting process. Because of the component material balances, anything that affects the composition on one end of the tower must have some effect on the composition at the other end.

Distillation is subject to constraints. If column optimization is being undertaken, various constraints must be included in the formulation.

7.8.1. MPC Formulation

Although MPC can output directly to final control elements, the manipulated variables for MPC are preferably the set points to flow controllers for D, B, L, and S.

MPC has the potential for ascertaining the best approach for controlling a tower and then proceeding to do just that. However, there is a catch. To make this option available to MPC for the column in Figure 7.1, the multivariable control problem must be defined as follows:

Manipulated Variable	Controlled Variable
Distillate flow set point D_{SP}	Distillate composition y_H
Bottoms flow set point B_{SP}	Bottoms composition x_L
Reflux flow set point L_{SP}	Reflux drum level H_D
Reboiler steam flow set point S_{SP}	Bottoms level H_B

If desired, pressure control could be included to make this a 5×5 multivariable process, but column pressure is generally controlled by manipulating condenser cooling

However, the temptation is to define the multivariable control problem as follows:

Manipulated Variable	Controlled Variable
Reflux flow set point L_{SP}	Distillate composition y_H
Reboiler steam flow set point S_{SP}	Bottoms composition x_L

Unfortunately, this restricts MPC to effectively using the L,V control configuration in Figure 7.2. MPC is better at coping with the interaction than individual PID controllers. However, the lesser the degree of interaction, the better MPC can perform.

The usual formulation of MPC is based on nonintegrating processes. The two composition processes are nonintegrating. However, the two-level processes are integrating. MPC can be formulated for integrating processes, and most commercial packages support integrating processes as well as noninte-grating processes.

7.8.2. Linear Systems Theory

MPC relies on finite response models and the principle of superposition from linear systems theory. Distillation is a nonlinear process, so such assumptions are valid only for operating within a region close to the base case. MPC requires extensive process testing, which must also be done within a region close to the base case. The MPC formulation includes parameters to make the controller more "robust," which means it is more tolerant of errors in the model. However, this comes at a cost—the MPC controller responds more slowly. In this regard, making MPC more robust is analogous to decreasing the controller gain in a PID controller.

The previous section illustrated that a major shift in the operating objec-tives for a column could require that the control configuration be changed—just retuning the controllers is not always adequate. What would be the implication of a major shift in the operating objectives on MPC? If the MPC is based on the test data from the original operating region, MPC would be taking control actions based on the original process behavior. The process behavior for the new operating objectives is different—and if a different loop configuration is required, the behavior must be very different. For major shifts in process behavior, the process tests must be repeated so that the control actions taken by MPC are consistent with the behavior of the process for the new operating objectives.

8

COMPLEX TOWERS

As used herein, a "complex tower" is anything beyond a simple two-product tower. The possibilities include

- heat integration,
- side heater and/or side cooler, and
- one or more sidestreams.

Occasionally, these are combined in complex fractionators that seem to have one common attribute: The feed is a naturally occurring material that the column splits into a number of product streams. Two common examples of such columns are the following:

Crude still (oil refining). A crude still splits the crude oil feed into a multitude of product streams, most of which are processed further to produce various final products.

Tall oil fractionator (paper industry). A tall oil fractionator splits the resin recovered from the digesters in the paper industry into a range of oils and waxes.

The complexity of these units is substantial. The tall oil fractionator is a steam still that is operated under vacuum. The crude still is an atmospheric

---

Distillation Control: An Engineering Perspective, First Edition. Cecil L. Smith.
© 2012 John Wiley & Sons, Inc. Published 2012 by John Wiley & Sons, Inc.

fractionator that also uses steam. The technology pertaining to such fraction-ators is complex and very specific to these industries. Such columns are beyond the scope of this book.

8.1. HEAT INTEGRATION

When the cost of energy increased in the 1970s, distillation attracted consider-able attention. The increased energy costs provided an incentive for the process designers to examine various possibilities for reducing the requirements for utilities.

Within a tower, an economizer can be installed to recover some energy from the bottoms product and return this energy to the tower with the feed. Econo-mizers are in the realm of energy conservation, but are not in the realm of heat integration. This term is usually applied where two or more towers are somehow involved in the energy conservation effort.

Figure 8.1 is an example of heat integration that involves two columns. The condenser for the first column also serves as the reboiler to the second column. Otherwise, these towers are completely independent.

From a control perspective, the issue with heat integration is the potential loss of one or more degrees of freedom. One consequence of the design in Figure 8.1 is that a degree of freedom is lost. In addition to the equations that apply to the individual towers, the following equation must be satisfied:

$$Q_{C,1} = Q_{R,2},$$

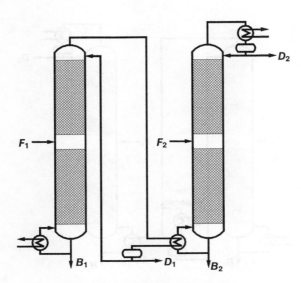

Figure 8.1. Heat integration.

where

$Q_{C,1}$ = heat removed in the condenser for tower 1;
$Q_{R,2}$ = heat added in the reboiler for tower 2.

If two towers are completely independent, it is possible to implement double-end composition control in both towers. But with the heat integration illustrated in Figure 8.1, this is no longer possible. Double-end composition control can be implemented in one tower, but only single-end composition control can be implemented in the other.

8.1.1. Trim Condenser

One approach to recovering the degree of freedom is to add a trim condenser as in Figure 8.2. The trim condenser permits more heat to be removed from column 1 than is added to column 2. This is subject to the following inequality:

$$Q_{C,1} \leq Q_{R,2} \leq Q_{C,1} + Q_{C,\text{Trim}},$$

where

$Q_{C,\text{Trim}}$ = maximum heat that can be removed by the trim condenser.

This recovers the degree of freedom, but only within what is often a narrow range. For cost reasons, the trim condenser is usually small.

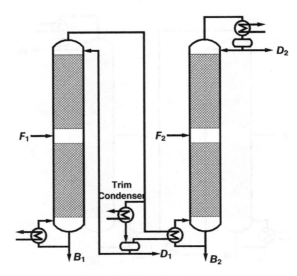

Figure 8.2. Trim condenser.

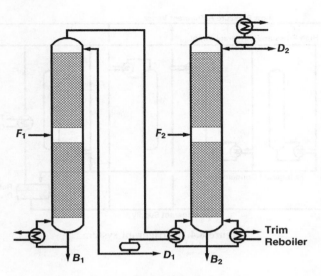

Figure 8.3. Trim reboiler.

8.1.2. Trim Reboiler

Another approach to recovering the degree of freedom is to add a trim reboiler as in Figure 8.3. The trim reboiler permits more heat to be added to column 2 than is removed from column 1. This is subject to the following inequality:

$$Q_{R,2} \leq Q_{C,1} \leq Q_{R,2} + Q_{R,\text{Trim}},$$

where

$Q_{R,\text{Trim}}$ = maximum heat that can be added by the trim reboiler.

Like trim condensers, trim reboilers are usually small, so the degree of freedom is recovered only within a narrow range.

8.1.3. Making the Case

There is no technical reason why both a trim condenser and a trim reboiler cannot be installed. The obstacle is cost. Even getting one installed can be difficult.

Process designers are not always sympathetic to issues pertaining to degrees of freedom, observing that "the material balances close, the energy balances close, so just operate the plant where it is designed to operate." Often issues pertaining to plant startup can be more effectively used to justify either a trim reboiler or a trim condenser. At least this levels the playing field—those

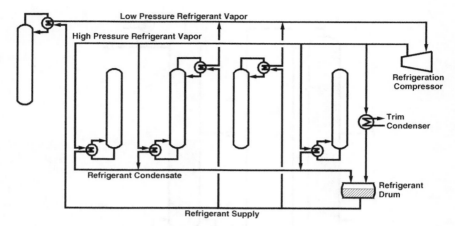

Figure 8.4. Refrigerant system.

computer printouts of steady-state material and energy balances are irrelevant for discussions pertaining to startup.

In the 1970s, some of the designs for heat integration were very ambitious. Fortunately, it did not take too long for some sanity to return. Incorporating heat integration into a process design tends to lock the process into the operating conditions for which it was designed. The problem is not the design, but the fact that for unforeseen reasons, it may be necessary to operate the process at conditions other than those for which it was designed.

8.1.4. Refrigeration Systems

Another potential source of a constraint is in shared equipment such as the refrigeration system illustrated in Figure 8.4. The system provides the following:

- The hot gas from the refrigeration compressor supplies heat to the reboilers for several columns. Basically, a reboiler serves as a condenser for the refrigerant.
- The liquid refrigerant is the coolant for the condensers for several columns. Basically, a condenser serves as an evaporator for the refrigerant.

The configuration in Figure 8.4 includes a trim condenser in the refrigerant system. Without the trim condenser, the refrigerant system imposes a constraint on process operations: the sum of the refrigerant vaporized in the column condensers must equal the sum of the refrigerant condensed in the column reboilers.

The trim condenser is normally recognized as being necessary during startup. But occasionally, someone proposes to shut down the trim condenser during normal production operations. If this is done, the constraint described above comes into effect.

8.2. SIDE HEATER/SIDE COOLER

Especially in chemical towers separating highly nonideal mixtures, large changes can occur in the vapor and liquid flows within a separation section. The effect of a side heater is as follows:

- Increase the vapor flow above the side heater.
- Decrease the liquid flow below the side heater.

The purpose of a side cooler is the opposite:

- Decrease the vapor flow above the side heater.
- Increase the liquid flow below the side heater.

In a tray tower, a side heater or side cooler can be added at any stage. In a packed tower, a side heater or side cooler can only be added between two packed sections.

8.2.1. Side Heater

The purpose of the side heater (sometimes called an interreboiler) illustrated in Figure 8.5 is to add heat to a stage. The net result of the side heater is to vaporize liquid.

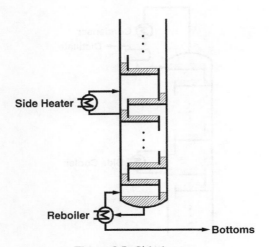

Figure 8.5. Side heater.

The heating media for side heaters can be steam, hot oil, or another process stream. The liquid is normally pumped through the side heater so that the exchanger can be located at grade level. In Figure 8.5, the partially vaporized stream leaving the side heater is returned to the stage from which the liquid was withdrawn. However, it is sometimes returned to another stage.

The desire is to vaporize a constant amount of liquid on the stage. When steam is the heating media, this is essentially accomplished by controlling the steam flow to the side heater. With hot oil, the heat transfer rate must be computed from hot oil flow, hot oil supply temperature, and hot oil exit temperature. The hot oil flow is then adjusted to give the desired heat transfer rate. When the heating media is another process stream, a bypass must be provided, normally on the process stream. Conceptually, the heat transfer rate can be computed in the same manner as for the hot oil. But if the process stream is a mixture such as a petroleum fraction, the liquid heat capacity may not be accurately known.

8.2.2. Side Cooler

The purpose of the side cooler (sometimes called an intercooler) illustrated in Figure 8.6 is to remove heat from a stage. The net result of the side cooler is to condense vapor.

Side coolers can be water-cooled, can be air-cooled, or can exchange heat with another process stream. The liquid is normally pumped so that the exchanger can be located at grade level. All of the liquid can be withdrawn (as illustrated in Fig. 8.6), or only part of the liquid can be withdrawn. Another variation, called a pumparound, withdraws liquid from one stage, pumps it

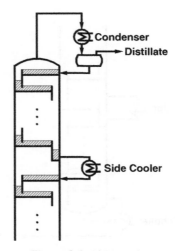

Figure 8.6. Side cooler.

through an exchanger to remove heat, and then returns the liquid to the tower at a few stages above the one from which it was withdrawn.

8.2.3. Control Configurations

Herein only control configurations for a side cooler will be presented. The configurations assume that liquid is withdrawn from the tower, pumped through an external exchanger, and then returned to the tower. Two configurations will be presented:

- Control liquid return temperature only.
- Control liquid return temperature and liquid flow.

For each configuration presented, an analogous configuration can be applied to a side heater.

8.2.4. Control Liquid Return Temperature Only

One way to vary the heat removed in a side cooler is to adjust the liquid return temperature. Maintaining a constant liquid return temperature removes a constant amount of heat provided the following are constant:

- Flow through the pump.
- Liquid heat capacity.
- Side cooler liquid inlet temperature (same as the temperature of the stage from which liquid is withdrawn). If this temperature varies, consider controlling to a fixed temperature change from the side cooler liquid inlet to the liquid return temperature.

One possibility is to manipulate a variable on the media side. A previous chapter discussed issues that arise for the tower condenser. The issues for a side cooler are the same. For a side heater, the issues are the same as discussed for reboilers in a previous chapter.

Providing a bypass around the side cooler as illustrated in Figure 8.7 permits the rate of heat transfer to be varied via a manipulated variable on the process side. The configuration in Figure 8.7 provides two control valves, one in the bypass and one on the side cooler exit. A three-way valve is an alternative, but the cost is about the same. Installing a control valve only in the bypass reduces costs, but does not permit all of the liquid to bypass the exchanger.

When two valves are provided as in Figure 8.7, the flow through the pump should be restricted as little as possible. Consider the following split range configuration (also illustrated in Fig. 8.7):

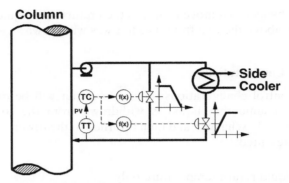

Figure 8.7. Control liquid return temperature only.

TC Output (%)	Bypass Valve Opening (%)	Exchanger Valve Opening (%)	Condition
0	100	0	No cooling
50	100	100	Intermediate cooling
100	0	100	Full cooling

At midrange, both control valves are fully open. The exchanger valve closes below midrange; the bypass valve closes above midrange.

The liquid return temperature controller in Figure 8.7 is sometimes the inner loop of a cascade, with the outer loop being a stage temperature loop, a product composition loop, or otherwise. For a side heater, the outer loop is sometimes the differential pressure across a separation section or some number of stages.

8.2.5. Control Liquid Return Temperature and Liquid Flow

In some towers, the designers specify the temperature at which the liquid is to be returned to the tower. This may be a fixed value, or may be a specified difference from the temperature of the stage where the liquid is returned.

How does one vary the heat removed in the side cooler? By varying the flow through the side cooler. The flow controller is often the inner loop of a composition or temperature cascade.

Controlling liquid return temperature and liquid flow entails two controllers:

- Liquid return temperature controller
- Liquid flow controller

There are two manipulated variables:

- Exchanger valve opening
- Bypass valve opening

The result is a 2×2 interacting process. Opening the bypass valve increases both the liquid return temperature and the liquid flow. Opening the exchanger valve decreases the liquid return temperature but increases the liquid flow.

Applying the relative gain to assess the interaction between the two loops suggests that the liquid flow should be controlled by manipulating the control valve with the larger flow. But there are possible complications:

- What if the flow is about evenly split between the bypass and the exchanger?
- What if at times most of the flow is through the bypass but at other times most of the flow is through the exchanger?

The control configuration in Figure 8.8 includes two summers that compensate for the interaction in the process. As a result, the actions taken by the controllers are as follows:

- On increasing its output, the liquid flow controller increases the opening of both control valves. The effect on the liquid return temperature should be minimal.
- On increasing its output, the liquid return temperature controller increases the opening of the exchanger valve but decreases the opening of the bypass valve. The effect on the liquid flow should be minimal.

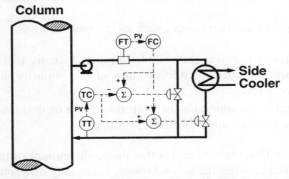

Figure 8.8. Control liquid flow and liquid return temperature.

8.3. SIDESTREAMS

Towers with a single sidestream are relatively common. Such towers produce three products:

- Distillate
- Sidestream
- Bottoms

Although not as common, columns can have multiple sidestreams. The more sidestreams, the more complex the tower. Such towers are also likely to have side heaters and/or side coolers, and some even have multiple feed streams.

A sidestream may be either of the following:

Liquid sidestream. Liquid sidestreams are withdrawn from above the feed stage.

Vapor sidestream. Vapor sidestreams are withdrawn from below the feed stage.

For tray towers, a sidestream may be withdrawn from any stage. For packed towers, a sidestream may only be withdrawn between two packed sections.

8.3.1. Use of a Sidestream

Sidestream towers are commonly installed in applications with the following requirements:

- A process stream contains a small amount of a volatile impurity that must be removed. Such impurities are often contaminants such as H_2S that must be reduced to a very low level.
- The remainder of the process stream is separated into two products similar to the separation provided by a two-product column.

One approach is to use two towers, such as in Figure 8.9:

- The first tower removes the volatile impurity. Usually, this is a relatively easy separation, but the concentration of the impurity in the bottoms must be very low.
- The second tower separates the bottoms stream from the first tower into a distillate product and a bottoms product.

The specification that must be met is the allowable amount of the impurity in the distillate product from the second tower. There may be a similar specification for the bottoms product, but this is unlikely to be exceeded.

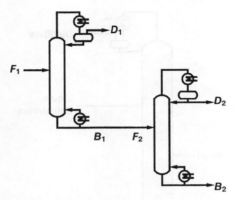

Figure 8.9. Two towers in series.

8.3.2. Controlled Variables for Two Towers

With two separate towers, one can control four compositions (two in each tower):

- The contaminant (H_2S) in the bottoms stream from tower #1. Usually, this contaminant is the light key.
- The total amount of organic material in the distillate from tower #1. Organics in this stream are usually lost.
- The light key in the bottoms stream from tower #2. This depends on the materials being separated.
- The heavy key in the distillate stream from tower #2. This depends on the materials being separated.

For contaminants such as H_2S, the specification is usually on the total sulfur in the distillate product from tower #2. In order to meet this specification, the amount of H_2S in the bottoms from tower #1 must be sufficiently low. Essentially, all H_2S in the feed to tower #2 leaves with the distillate product.

8.3.3. Tower with Liquid Sidestream

Instead of two towers, most process designs would meet these requirements with a single tower with a sidestream. One way to view the tower in Figure 8.10 is as follows:

- The main purpose of the tower is to separate the feed into the sidestream product and the bottoms product. The sidestream product is the counterpart to the distillate product from tower #2 in the two-tower configuration in Figure 8.9. This split is achieved with the separation section below the

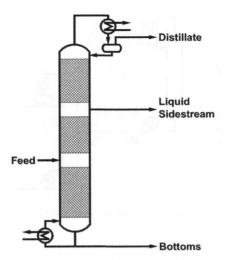

Figure 8.10. Liquid sidestream.

feed stage and the first separation section above the feed stage. These two separation sections will be almost the same as the two separation sections of tower #2 of the two-tower configuration.

- The contaminant H_2S occurs in minor amounts and is removed via the distillate stream. The purpose of the upper separation section is to concentrate the H_2S in the tower vapor stream, thus reducing the loss of organics with the distillate stream.

The sidestream tower is basically viewed as tower #2 in the two-tower configuration, but with an extra separation section to remove the contaminant through the distillate stream. Because materials like H_2S are difficult to condense, the distillate stream is often a vapor stream. However, it is usually too small for effective pressure control.

8.3.4. Controlled Variables for Sidestream Tower

For a tower with one sidestream, a maximum of three compositions can be controlled. Consider the light key and the heavy key in the context of separating the feed into the sidestream product and the bottoms product. For this application, the following compositions would be controlled:

- The composition of the light key (LK_B) in the bottoms product.
- The composition of the heavy key (HK_{SS}) in the sidestream product.
- The composition of the contaminant (S_{SS}) in the sidestream product.

The distillate is a vapor stream and is usually a small flow relative to the other flows. Increasing the distillate flow reduces the concentration of the contaminant in the sidestream. Consequently, the composition of the contaminant in the sidestream is usually controlled by adjusting the distillate flow.

Often the distillate flow is adjusted manually and in a very conservative manner; that is, the distillate flow is much larger than necessary. But this incurs a cost—the loss of organics with the distillate product is higher than necessary.

8.4. WITHDRAWING A LIQUID SIDESTREAM

From the perspective of separation within the column, the main issue is the internal liquid flow L_I to the separation section immediately below the liquid sidestream. If the vapor and liquid flows within the upper separation section are constant (i.e., equimolal overflow), the internal liquid flow can be computed as follows:

$$L_I = L - SS.$$

But given the nature of the materials being separated by a tower with a sidestream, equimolal overflow is unlikely. That means that the liquid flow out of the upper separation section will not be the same as the liquid flow into the upper separation section.

The factor by which the liquid flow increases or decreases within the upper separation section could be computed from the solution of the stage-by-stage separation model. But the accuracy of this factor is questionable. Nonideal materials usually exhibit large changes in the liquid and vapor flows within a separation section. Errors in the data on the thermodynamic properties for such materials lead to errors in the factor for the change in liquid flow.

8.4.1. Example

The following example illustrates the type of situation that can be encountered in towers with a liquid sidestream:

- The external reflux L to the top of the tower is 50 units.
- The liquid sidestream SS is 80 units.
- The desired liquid flow L_I immediately below the sidestream is 20 units (the minimum liquid rate required to wet the packing).

In this tower, the liquid flow within the upper separation section increases by a factor of 2, giving 100 units of liquid out of the upper separation section.

Of this, 80% must be withdrawn at the sidestream, leaving 20% below the sidestream. How can one reliably do this? Two large numbers (100 units and

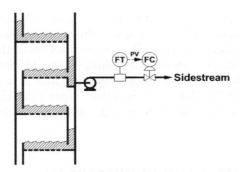

Figure 8.11. Partial withdrawal of liquid leaving a stage.

80 units) are being subtracted to give a small one (20 units). Variations in the big numbers will be amplified in the small one. The liquid flow below the sidestream is the minimum required to wet the packing. Any errors in this flow on the low side have adverse consequences.

8.4.2. Partial Withdrawal of Liquid

Special internals permit part of the liquid from a stage to be withdrawn, with the extra liquid overflowing to the stage below. The internals illustrated in Figure 8.11 are for a tray tower. For a packed tower, appropriate internals are available for withdrawing a liquid stream between packed sections.

The amount of liquid withdrawn at the sidestream can be controlled very accurately. This must be considered in light of the approach to controlling the composition of the light key in the sidestream:

Composition is controlled by manipulating the sidestream flow (the direct material balance approach). The ability to accurately control the flow of the sidestream is all that is required.

Composition is controlled by manipulating the reflux flow below the sidestream (the indirect material balance approach). The reflux flow below the sidestream is the liquid flowing from the upper separation section less what is withdrawn at the sidestream. This flow cannot be accurately computed or controlled.

Those focused on costs find the partial withdrawal of liquid from a stage very appealing. All other alternatives are more costly.

8.4.3. Total Withdrawal from Internal Reservoir

In the tower illustrated in Figure 8.12, all of the liquid from the upper separation section flows into an internal reservoir. All of the liquid is removed from

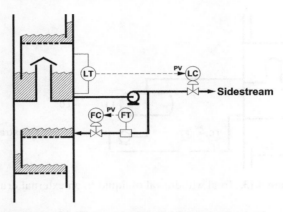

Figure 8.12. Total withdrawal of liquid from the internal reservoir.

the tower. Some goes to the sidestream; the remainder is returned to the tower as reflux to the separation section below the sidestream.

A level transmitter is provided for the internal reservoir. This level can be controlled either with the sidestream or with the reflux to the tower:

Control level by manipulating sidestream flow. As illustrated in Figure 8.12, the control loops are as follows:

- A flow controller is provided for the reflux flow. This loop is often the inner loop of a composition or temperature cascade.
- The level in the internal reservoir is controlled by manipulating the sidestream flow.

With this configuration, the reflux flow below the sidestream is both known (it is directly measured) and can be accurately controlled. Neither is possible for the partial withdrawal of liquid arrangement in Figure 8.11.

Control level by manipulating reflux flow. The control loops are as follows:

- A flow controller is provided for the sidestream flow. This loop is often the inner loop of a composition or temperature cascade.
- The level in the internal reservoir is controlled by manipulating the reflux returned to the tower.

With this configuration, the variability in the reflux flow below the sidestream would be comparable with that of the partial withdrawal arrangement in Figure 8.11. But there is an advantage: The reflux flow can be measured and a minimum imposed on the allowable reflux flow.

As compared with the partial withdrawal of liquid in Figure 8.11, the arrangement in Figure 8.12 is more costly. The obvious costs are the level measurement and the additional control valve. However, the presence of the liquid

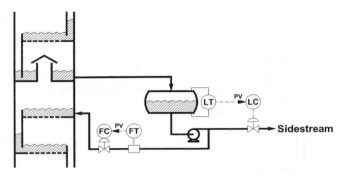

Figure 8.13. Total withdrawal of liquid to an external drum.

reservoir adds considerable weight at a point far up the tower. This could raise structural issues in towers without an external structure for support. Small-diameter chemical towers usually require an external structure for support, so the structural issues resulting from the additional weight can be addressed relatively easily.

8.4.4. Total Withdrawal to an External Drum

In the tower illustrated in Figure 8.13, the tower internal (a chimney tray) collects all of the liquid and directs it to an external reflux drum. Some goes to the sidestream; the remainder is returned to the tower as reflux to the separation section below the sidestream.

In towers that do not require an external structure for support, the external drum can be located at grade level, which usually minimizes the structural issues. Also, the level transmitter is easily accessible instead of being physically located near the top of the tower. For the small-diameter chemical towers that require a structure for support, these advantages are not significant.

The structural issues usually determine whether an internal reservoir as in Figure 8.12 or an external reflux drum as in Figure 8.13 will be installed. Both approaches permit the reflux flow to the tower to be measured and controlled. Since the sidestream flow can also be measured, these configurations also permit either of the following ratio configurations to be implemented:

- Ratio reflux flow to sidestream flow.
- Ratio sidestream flow to reflux flow.

8.5. WITHDRAWING A VAPOR SIDESTREAM

From the perspective of separation within the column, the main issue is the internal vapor flow V_I to the separation section immediately above the sidestream. Computing the vapor flow above the sidestream involves assumptions such as equimolal overflow, which are unlikely to be valid for the materials usually being separated by a tower with a sidestream. Basically, the issues are the same as previously discussed for towers with a liquid sidestream.

8.5.1. Partial Withdrawal of Vapor

The desire is to withdraw a vapor sidestream free of any entrained liquid. This can generally be done between the trays of a tray tower or between the packed sections of a packed tower.

The amount of vapor withdrawn at the sidestream can be controlled very accurately. This must be considered in light of the approach to controlling the composition of the light key in the sidestream:

Composition is controlled by manipulating the sidestream flow. The ability to accurately control the flow of the sidestream is all that is required.

Composition is controlled by manipulating the vapor flow above the sidestream. The vapor flow above the sidestream is the vapor flowing from the lower separation section less what is withdrawn at the sidestream. This flow cannot be accurately computed or controlled.

For vapor sidestreams, there is no counterpart to the configurations in Figure 8.12 or Figure 8.13 for a liquid sidestream. There is no practical way to withdraw all of the vapor, return a specified amount to the tower, and permit the remainder to exit as the sidestream.

8.5.2. Controlling Column Differential Pressure

In those cases where the vapor flow above the sidestream is of primary interest, the configuration illustrated in Figure 8.14 should be considered:

- Measure the differential pressure across the separation section immediately above the sidestream.
- Control this differential pressure by adjusting the vapor sidestream flow. If the differential pressure is increasing, the controller must increase the vapor sidestream flow. The configuration in Figure 8.14 is a differential-pressure-to-flow cascade.

Especially when the vapor flow above the sidestream approaches the limit for tower flooding, this approach generally works well. But if the tower is operating well below its flooding limit, this approach suffers from the same problem as head-type flow meters at low flow rates. The differential pressure varies with the square of the flow (or actually vapor velocity). At low flow rates, the change

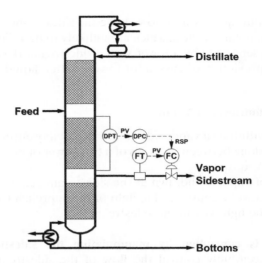

Figure 8.14. Vapor sidestream.

in differential pressure due to the change in the vapor flow is too small to be usable. Even linearizing the relationship by taking the square root is of little help. At low flows, taking the square root amplifies the noise in the differential pressure measurement.

8.6. COMPOSITION CONTROL IN SIDESTREAM TOWERS

Figure 8.15 illustrates a tower with a liquid sidestream that is in the service described previously. The feed contains a minor amount of a very volatile contaminant that is removed via the distillate stream. The remainder of the feed is split into the sidestream product and bottoms product. This service is typical of many, but not all, towers with a liquid sidestream.

Such towers can basically be viewed as a two-product tower (the sidestream and the bottoms) that makes a split between the light key component and the heavy key component. The distillate stream is a small vapor stream whose flow is adjusted to remove the contaminant. Using such a perspective, control configurations for two-product towers can be extended to towers such as in Figure 8.15.

8.6.1. Controlled and Manipulated Variables

The tower in Figure 8.15 has an external reflux drum for the sidestream. This gives a total of seven controlled variables and seven manipulated variables:

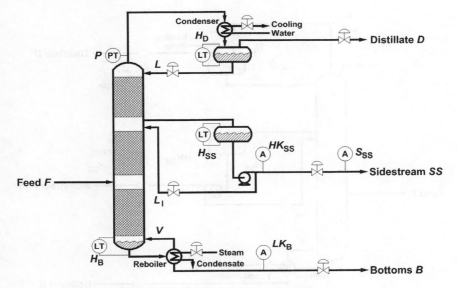

Figure 8.15. Tower with liquid sidestream.

Controlled Variable	Manipulated Variable	Control Valve
Distillate reflux drum level H_D	Distillate flow D	Distillate
Sidestream reflux drum level H_{SS}	Sidestream flow SS	Sidestream
Bottoms level H_B	Bottoms flow B	Bottoms
Column pressure P	Reflux flow to top of tower L	Reflux
Contaminant in sidestream S_{SS}	Reflux flow below sidestream L_I	Sidestream reflux
Heavy key in sidestream HK_{SS}	Heat removed in condenser Q_C	Cooling water
Light key in bottoms LK_B	Heat added in reboiler Q_R	Steam

There is no significance to the order of the controlled and manipulated variables in the above list.

8.6.2. Contaminant Composition Control

As the contaminant is very volatile, the distillate product is a vapor stream. However, the flow is too small to be used to control the column pressure. Instead, the distillate flow is manipulated to control the concentration of the contaminant in the sidestream.

In the configuration in Figure 8.16, the contaminant composition controller manipulates the set point to the distillate flow controller. However, in many towers, this loop is not on closed-loop control. Instead, the operators manually adjust the set point of the distillate flow controller to maintain the contaminant level in the sidestream well below the limit imposed by the specifications.

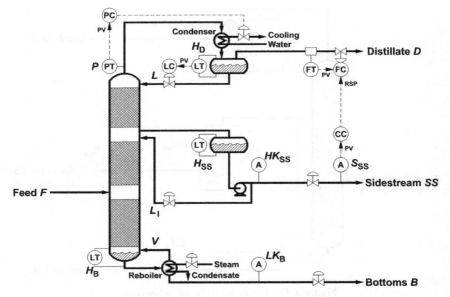

Figure 8.16. Controlling the concentration of contaminant in sidestream.

Manipulating the distillate flow to control contaminant concentration dictates the following two loops:

- The column pressure P is controlled by manipulating the heat removed in the condenser Q_C.
- The top reflux drum level H_D is controlled by manipulating the reflux flow L to the top of the tower.

8.6.3. Control of Sidestream Composition and Bottoms Composition

With the column pressure, top reflux drum level, and sidestream contaminant concentration being controlled using the configuration in Figure 8.16, the remaining controlled and manipulated variables are as follows:

Controlled Variable	Manipulated Variable	Control Valve
Sidestream reflux drum level H_{SS}	Sidestream flow SS	Sidestream
Bottoms level H_B	Bottoms flow B	Bottoms
Heavy key in sidestream HK_{SS}	Heat removed in condenser Q_C	Cooling water
Light key in bottoms LK_B	Heat added in reboiler Q_R	Steam

This list is identical to the list for double-end composition control in a two-product tower. All of the control configurations applicable to double-end composition control are applicable to the sidestream tower in Figure 8.15 subject to the following equivalents:

Two-Product Tower	Tower with Liquid Sidestream
Distillate flow D	Sidestream flow SS
Reflux flow L	Reflux flow below sidestream L_I

8.6.4. D,V Configuration

Using the counterpart of the D,V configuration for a two-product tower, the compositions are controlled as illustrated in Figure 8.17:

Controlled Variable	Manipulated Variable
Heavy key in sidestream HK_{SS}	Sidestream flow SS
Light key in bottoms LK_B	Boilup flow V

With this configuration for controlling the compositions, the levels are controlled as follows:

- The sidestream reflux drum level H_{SS} is controlled by manipulating the reflux flow below the sidestream L_I.
- The bottoms level H_B is controlled by manipulating the bottoms flow B.

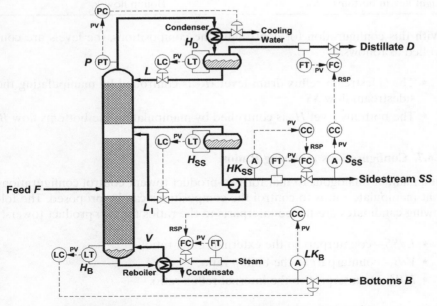

Figure 8.17. Sidestream tower control via D,V configuration.

8.6.5. *L,B* Configuration

Using the counterpart of the *L,B* configuration for a two-product tower, the compositions are controlled as follows:

Controlled Variable	Manipulated Variable
Heavy key in sidestream HK_{SS}	Reflux flow below sidestream L_I
Light key in bottoms LK_B	Bottoms flow B

With this configuration for controlling the compositions, the levels are controlled as follows:

- The sidestream reflux drum level H_{SS} is controlled by manipulating the sidestream flow SS.
- The bottoms level H_B is controlled by manipulating the boilup V.

8.6.6. *L,V* Configuration

Using the counterpart of the *L,V* configuration for a two-product tower, the compositions are controlled as follows:

Controlled Variable	Manipulated Variable
Heavy key in sidestream HK_{SS}	Reflux flow below sidestream L_I
Light key in bottoms LK_B	Boilup flow V

With this configuration for controlling the compositions, the levels are controlled as follows:

- The sidestream reflux drum level H_{SS} is controlled by manipulating the sidestream flow SS.
- The bottoms level H_B is controlled by manipulating the bottoms flow B.

8.6.7. Configurations Involving Ratios

In a manner analogous to that for two-product towers, control configurations that manipulate ratios to control the compositions can be proposed. The following candidates are the counterparts to the ratios for two product towers:

- L_I/SS—counterpart to the external reflux ratio;
- V/B—counterpart to the boilup ratio;
- L_I/V—counterpart to the internal reflux ratio.

However, several more ratios could be proposed for a sidestream tower.

INDEX